bluespec

BSV by Example

The next-generation language for electronic system design

Rishiyur S. Nikhil and Kathy R. Czeck

2

Edition 1.0

ISBN: 978-1456418465

Any questions may be directed to support@bluespec.com. Additional support resources are available at http://www.bluespec.com/support/.

The authors thank all our colleagues and friends at Bluespec, MIT and Sandburst who contributed to the creation of BSV, its tools, and to the expertise for their effective use. In this book, the authors are merely channeling this vast body of shared knowledge.

Contents

Chapter 1

Introduction

BSV (Bluespec SystemVerilog) is a language used in the design of electronic systems (ASICs, FPGAs and systems). BSV is used across the spectrum of applications—processors, memory subsystems, interconnects, DMAs and data movers, multimedia and communication I/O devices, multimedia and communication codecs and processors, signal processing accelerators, high-performance computing accelerators, etc. BSV is also used across markets—from low-power, portable consumer items to enterprise-class server-room systems.

Being both a very high-level language as well as fully synthesizable to hardware, it is used in many design activities, as described below. This combination of high level and full synthesizability enables many of these activities that were previously only done in software simulation now to be moved easily to FPGA-based execution (whether the design is eventually bound for ASICs or FPGAs). This can speed up execution by three to six orders of magnitude (1000x to 1,000,000x). Such a dramatic speedup not only accelerates existing activities, but enables new activities that were simply not feasible before (such as cycle-accurate multi-core processor architecture models running real operating systems and real applications).

1.1 Design activities where BSV is used

Executable Specifications (synthesizable). For today's complex systems, a specification written in a human language (like English) is likely to be imprecise, incomplete or even infeasible and self-contradictory in its requirements. A specification in BSV addresses these concerns, because of its precise semantics and executability on real data. Fleshing out a spec in this way is also called *spec validation.*

Virtual Platforms (synthesizable). Today's chips are dominated by the volume and complexity of the software that runs on them. It is no longer acceptable to have to wait for chips to be available before developing software. Virtual Platforms enable software development and testing to begin as early as day one. Virtual platforms written in BSV, being synthesizable, run on FPGAs at much higher speeds and greater accuracies than traditional software virtual platforms, greatly increasing software development productivity.

Architectural Modeling (synthesizable). Complex SoCs (Systems on a Chip) involve complex trade-offs. What should be done in SW vs. in HW accelerators? How many processors vs. HW accelerators? What micro-architecture for processors, interconnect, memory systems? These decisions will affect chip area, speed, power consumption, cost, and time-to-market. Many of these decisions are too complex to be taken analytically—they require actual execution on actual data. BSV is used for synthesizable architecture models, exploration and validation at the system, subsystem, and the IP block levels.

Design and Implementation. This is the classic design activity of IP blocks (Intellectual Property), except that today these can be complex subsystems. BSV enables designing such components at a much higher level of abstraction and with better maintainability.

Verification Environments (synthesizable). Every verification environment is a model of "the rest of the system." As such, they can be as complex as designs themselves, face similar speed-of-execution issues, and face similar issues of reusability, maintainability and evolvability. BSV is used for writing synthesizable transactors, testbenches and both system and reference models.

1.2   Key ideas in BSV

BSV borrows powerful ideas that have been developed over several decades in advanced formal specification and programming languages. These can be classified into two groups—behavior and structure.

1.2.1   Behavior

BSV's behavioral model is popular among formal specification languages for complex concurrent systems. We call them *Atomic Rules and Interfaces*, but in the literature they also go by various names such as Term Rewriting Systems, Guarded Atomic Actions, and Rewrite Rules. Well-known formal specification languages that use a similar model include Guarded Commands (Dijkstra); TLA+ (Lamport); UNITY (Chandy and Mishra); and Z, B and EventB (Abrial).

The reasons BSV and all these formal specification languages choose this model are:

- *Parallelism:* The model of atomic rules is fundamentally parallel, unlike many other approaches that graft a parallel extension on top of a sequential language (such as those based on C++). This is ideally suited for the massive, fine-grain, heterogeneous parallelism that is everywhere in complex hardware designs and distributed systems.

- *Correctness:* The atomicity of rules is a powerful tool in thinking about correct behavior in the presence of complex parallelism. A key concept in correct systems is the *invariant*, which represents a correctness condition that one expects to be maintained in a system (such as, "A cache line can be in the DIRTY state in only one of the multiple caches", or "The counter in the DMA is equal to the number of bytes transferred"). Atomicity of rules allows reasoning about invariants one rule at-at-time. Without atomicity, one has to worry about all possible interleavings of parallel activities, an exercise that quickly becomes too complex both for humans and for formal verification systems.

The hardware designer sees profound practical consequences in expressing behavior using Atomic Rules and Interfaces. Any interesting piece of hardware has control logic (multiplexers, enable signals, and so on). This control logic can become extremely complex. Further, control logic typically spans module boundaries, and manifests itself at module boundaries in ad-hoc protocols and assumptions on signalling. Without a rigorous semantics, designs tend to accumulate layer upon layer of such assumptions, often poorly documented (if at all) and poorly communicated between the designer of a module and the users of the module. This is what makes RTL design so brittle, because even the smallest change can violate one of these myriad control assumptions, resulting in race-conditions, dropped data, mis-sampled data, buffer overflows and underflows, and the like. Atomic Rules and Interfaces are the solution—they provide a rigorous semantics and a much higher level view of interactions between modules, and the complex control logic is automatically synthesized, correct by construction. Even more important: when the source code is changed, a simple recompilation will re-synthesize all this complex control logic, accommodating all the changes in a correct manner. This allows the hardware designer to focus on overall architecture structure, and leave the nitty-gritty detail of control logic to the compiler (which is normally designed and maintained by hand by the RTL designer).

1.2.2 Structure

Modern high-level programming languages have developed some powerful abstraction mechanisms for organizing and structuring large, complex, software systems. Unfortunately, until BSV, many of these capabilities have largely passed hardware design languages by, in part under the perhaps mistaken presumption that those abstraction mechanisms are "not relevant" for hardware design languages.

BSV builds on, and borrows ideas from, two sources for its structural abstractions:

- *SystemVerilog*, for modules and module hierarchy; separation of interfaces from modules; syntax for literals, scalars, expressions, blocks, loops; syntax for user-defined types (enums, structs, tagged unions, arrays), and so on. All these will be very familiar to people who know Verilog and SystemVerilog.

- *Haskell*, for more advanced types, parameterization and static elaboration. Haskell is a modern "pure functional programming" language that is recently seeing a resurgence both because it addresses software complexity and because it is a promising basis for attacking the "parallel programming software crisis" that has been precipitated by the recent advent of multi-core and multi-threaded processors. Haskell has many features that are well acknowledged to be more powerful than those in languages like C++, Java and C#.

1.3 Why BSV? Can't we do all this with existing languages (like C++)?

[This section is for those particularly interested in this topic, and can be skipped without loss of continuity.]

When someone comes to you with "a new language" it is often time to run the other way. Often, new languages offer slightly different features, repackaging of existing features, and so on, most of which perhaps could have been achieved with a suitable library or "style guide" for an existing language. The advantages of sticking to an existing language are of course tremendous, given their maturity and vast existing ecosystem of tools and expertise that one can draw upon.

Unless ... the new language offers such compelling value that it is worth the switch (and, we believe, this is the case with BSV). This value must be at least an order of magnitude (10x) to be worth the effort of changing horses.

In the software world, people switched from C to C++ because abstraction mechanisms such as object-orientation, stronger type-checking, templates, and overloading could not simply be grafted onto C via libraries or style guides. Others switched further to Java and C# because of automatic storage management, still more expressive types, portability and greater simplicity. New scripting languages (such as Tcl, Python and Ruby) emerged for more dynamic interactivity and behavior. (Interestingly, Python also borrows many ideas from Haskell.)

Today, in the hardware design space, people use different languages for different design activities:

- *Executable specs.* These are still not the norm. Specs are still mostly written in human languages (like English). Some analysis may be done in spreadsheets.

- *Virtual Platforms.* Today these are usually pure software simulators based on C++ and/or SystemC.

- *Architectural Modeling.* This used to be done with various *ad hoc* tools, but recently there is a trend towards the SystemC TLM 2.0 (Transaction Level Modeling) methodology. The methodology encompasses various levels of timing accuracy.

- *Design and Implementation.* This is still mostly done with RTL (Verilog and VHDL) and classical RTL synthesis. There are some small steps up in abstraction from Verilog to SystemVerilog. In a few cases, people are using so-called *High Level Synthesis* (HLS) starting from C++ source codes. In even fewer cases, people are using synthesis from SystemC (but this is typically not HLS, as explained in more detail below).

- *Verification Environments.* From languages like *e* and *Vera* there is a move towards SystemVerilog and methodologies like VMM and UVM. Others are beginning to use SystemC. All of them exploit similar concepts, like object-orientation and constrained random testing.

In general, except for design and implementation, all the other activities are done purely in software, because the languages used (full C++, SystemC, object-oriented capabilities of SystemVerilog) are not synthesizable. This raises a serious problem—you can either get speed, or accuracy, but not both.

For speed, one starts with the native speed of compiled C++. Software virtual platforms use sophisticated technology like just-in-time cross-compilation to deliver impressive execution speeds, for example booting an entire guest operating system in seconds. TLM models avoid the overheads of event-driven simulation (by sticking to SC_METHOD and avoiding SC_THREAD and

SC_CTHREAD). If they must use events, they play tricks like "temporal decoupling" to amortize the overhead.

While all these tricks are laudable (and perhaps adequate for some modeling purposes), unfortunately performance drops steeply as one adds almost *any* architectural detail, such as actual parallelism, interconnect contention, memory or I/O bottlenecks, cycle-accurate device models, and so on. Performance also drops steeply when one adds any instrumentation at all (which is necessary for performance modeling). Further, since the basic substrate is C++, it is extraordinarily difficult to achieve any speed up by exploiting multi-core hosts.

In summary, as you add architectural detail (for accuracy), you lose speed. Development time increases sharply, as programmers contort the code to apply tricks to restore speed.

One attempted solution is to use FPGA-based execution for the architecturally detailed component(s) of the system model. This is still a problem because of the logistical difficulties of dealing with multiple languages and paradigms for different components, because the software side may still run too slowly, because bandwidth constraints in communicating between the software and FPGA sides may anyway limit speed, and because of limited visibility into the FPGA-side components. Further, conventional approaches have problems in creating the FPGA-side components (see synthesis discussion below).

High-Level Synthesis (HLS) from C++ and SystemC is often held out by many in the industry as a potential way forward. Unfortunately there are serious problems in using C++ and SystemC as a starting point. A central issue is *architecture, architecture, architecture.*

It is well understood that the key to high-performance software is good *algorithms.* Further, good algorithms are designed by people with training and experience, and not by tools (a compiler only optimizes an algorithm we write, it doesn't improve the algorithm itself). Similarly, the key to high-performance hardware (measured as area, speed, power) is good *architecture.* Hence, it is crucial to give designers the maximum power to express architecture; all the rest can be left to tools.

Unfortunately, HLS from C++ does exactly the opposite—it obscures architecture. These HLS tools try to choose an architecture based on internal heuristics. Designers have indirect, second-order controls on this activity, expressed as "constraints" such as bill-of-materials limits for hardware resources and pragmas for loop transformations (unrolling, peeling, fusion, etc.). It is hard (and often involves guesswork) to "steer" these tools towards a good architecture.

HLS from C++ has another serious limitation. It is fundamentally an exercise in "automatic parallelization", i.e., in automatically transforming a sequential program into a parallel implementation. This problem has been well-studied since the days of the earliest vector supercomputers like the CDC 6600 and Cray 1, and works (and even then, only moderately) for "loop-and-array" codes. This is because the key technologies needed for automatic parallelization, namely dependence analysis, alias analysis, etc., on the CDFG (Control/Data Flow Graph) have been shown to be feasible only for simple loops, often statically bounded, working on arrays with simple (affine, linear) combinations of loop indexes. Such loop-and-array computations of course exist in a few components in an SoC, notably the signal-processing ones, but are absent in the majority of the SoC, which are more "control-oriented" (processors, memory systems, interconnects, DMAs and data movers, I/O peripherals, etc.). Even in the DSP space, note that modern DSP algorithms are also becoming more control-oriented in order to become more adaptive to noise (for signal processing) and actual content (for audio and video codecs).

Finally, HLS from C++ has yet another serious limitation. It is well known that, for the same computation, one may choose different algorithms depending on the *computation model* and associated *cost model* of the platform. This is why there are entire textbooks on parallel algorithms—they would be unnecessary if sequential algorithms were adequate. Unfortunately, C++ has a fixed computational model—the von Neumann model of sequential execution and large, flat (constant-cost) memory. Consequently, algorithms written in C++ may not be the best choice for hardware implementation. Often, one must "reverse engineer" the mathematical function from a particular C++ algorithm before coming up with a good hardware architecture. Of course, no HLS tool is going to do that for you.

HLS from SystemC is almost a misnomer. While it may incorporate HLS from C++ for the bodies of some IP blocks as described in the previous paragraphs, the rest of the design (blocks that are not "loop-and-array", and system integration) are essentially expressed at the RTL level of abstraction and uses classical RTL synthesis techniques.

With this discussion of the C++/SystemC/SystemVerilog state of the art behind us, we can describe why BSV delivers the order of magnitude improvements in productivity and capability promised at the beginning of this section.

First, unlike C++ and SystemC TLM, BSV is *architecturally transparent*. The designer, and not the tool, explicitly and directly expresses architecture. BSV's powerful types and static elaboration mechanisms provide great power to express architectural structures very succinctly and programmatically—but static elaboration is deterministic and predictable and there are no surprises. As a result, HLS from BSV typically allows you to converge to a high-quality design much faster than HLS from C++, where you have to try "steering" the tool towards a good architecture.

Second, unlike C++ and SystemVerilog, BSV has a superior behavioral semantics—Atomic Rules and Interfaces—which is a higher-level abstraction for concurrency and is much better suited to the task of describing the fine-grain, multi-rate, heterogeneous parallelism found in hardware systems.

Because of the above two properties, with BSV one can go from a truly abstract (mathematical) description of an algorithm directly towards a good architectural implementation, without the detour through C++ code that has at best obscured it and at worst chosen a poor concrete sequential algorithm. And unlike HLS from C++, HLS from BSV can seamlessly encompass and interact with complex control logic, interconnects, memory subsystems, and so on.

Third, BSV has much stronger parameterization, which directly affects everything: code size, code structure, code reuse, and code correctness. All types in BSV are first-class, that is, one can write functions taking any types as arguments and any types as results. Modules and functions can be parameterized by other modules and functions; functions can generate Rules and Interfaces, and so on. This results in an unprecedented level of static elaboration expressive power (Turing complete "generate").

Fourth, BSV has a much stronger data type system (including polymorphic types and user-defined overloading) with much stronger type-checking. This gives more expressive power with greater safety.

Fifth, BSV also performs many other kinds of strong static checking. Chief amongst these is a semantically well-founded notion of clock and reset domains. For example, accidentally crossing a clock domain boundary without a synchronizer is a thing of the past.

Sixth, and finally, programs using all the above features are synthesizable—there is no retraction into any weaker "synthesizable subset" as one sees in C++, SystemC and SystemVerilog.

As a result, BSV also solves the "multitude of disparate languages" problem. A single, unified language can be used for synthesizable executable specs, synthesizable (fast, scalable) virtual platforms, synthesizable (fast, scalable) architecture models, synthesizable design and implementation, and synthesizable test environments. All of these can be run on FPGA platforms, the modern way to do extensive verification.

1.4   About this book

This book is intended to be a gentle introduction to BSV. Even though all the key ideas have been tried and tested for decades in other specification and programming languages (as cited in previous sections), they are often new to people in the hardware and hardware systems design community and, like all things new, it takes some practice for them to feel natural.

For example, for most of us brought up on a strict diet of sequential programming languages (Fortan, C, C++, Java, C#, Perl, Tcl, Python, even Matlab), the idea of describing behavior using Atomic Rules and Interfaces initially seems strange. The assumption of the sequential von Neumann model has been so implicitly and subconsciously wired into our way of thinking that anything else initially seems strange. After a while, as one awakens to the sense of how natural Atomic Rules and Interfaces are for describing hardware, one undergoes an epiphanous flip and begins to ask: "Why would one ever describe hardware in any other way?"

Similarly, if one has never encountered polymorphic types, user-defined overloading, or object-orientation, it takes a little getting used to, but after a while one wonders how one ever lived without them! In this regard the C++ programmer has an advantage over the C and Verilog programmer, because C++ has these features, although not with synthesis, and not with type-checking as strong as in BSV.

One of the subtle but insidious consequences of our traditional sequential languages is that we play fast and loose with side-effects (state change), precisely because sequentiality makes it tractable to reason about side-effects. But side-effects and parallelism are a potentially deadly combination, unless carefully mediated by atomic transactions. BSV goes to extraordinary lengths to separate "pure" expressions from "side-effecting" (state-changing) expressions, codifying the difference in the data types and bringing the power of static type-checking to enforce this (see Action and Action-Value types). This is again one of those features which initially seems strange; after it has saved your design (by static type-checking!) a couple of times, you begin to wonder how you ever survived without it!

Even the seasoned C++ programmer is often astonished by the power of functional programming in BSV, where all types are first-class, and one can write arbitrary "generate" computations, passing modules, functions, rules, interfaces freely as arguments and results. Even more astonishing, these features can be used in synthesizable programs.

This book is not a language reference manual; for that, please see Bluespec's *BSV Reference Guide*. Nor is it about how to use the tools; for that, please see the *User Guide*. Nor is it particularly about good hardware design (good architecture), which is a subject that people should learn no matter what design language they use. The prototypical target audience for this book is the person who says: "I have an idea for a good architecture that I can sketch out for you on the whiteboard;

now please give me a language where I can focus on expressing that idea precisely, concisely, safely, synthesizably, and without getting lost in a million details."

This book is also not about overall system design methodology. The discussion in this chapter should give the reader a sense of our recommended design methodology, *design by refinement*. That is, we start by writing executable specifications of systems, and we gradually and incrementally add architectural detail until we have reached our performance targets (speed, area and power in the chosen target technology). All this is done in a single language, BSV, so that there is an enormous reuse of code during the refinement, particularly reuse of interfaces, testbenches, test data, models, verification IP, and so on. Further, we execute repeatedly and frequently, from day one, starting with the very earliest approximations of the executable spec, so that at all times we have a calibration about feasibility, correctness, completeness and cost. However, this book does not take you through this design-by-refinement process with a major design example (some examples along with whitepapers are included in the Bluespec software distribution). Here, you will find various sequences of small examples that demonstrate small incremental changes, giving a flavor of refinement.

This book tries to take you into the BSV language one small step at a time. Each section includes a complete, executable (and synthesizable) BSV program, and tries to focus on just one feature of the language. Please study the actual code for the example; the text in the corresponding section only highlights certain interesting features, and the actual example code often has more detail of interest. A few examples span multiple sections, with each successive section showing a small but interesting modification or alternative to the previous ones. The complete source codes are in the appendix of the book, but are also available in machine-readable form. We strongly encourage you actually to build and execute the examples and, even better, try some of your own modifications—there's no better way to learn than by doing.

Chapter 2

Getting started with BSV

2.1 A simple example

Source directory: `getting_started/hello_world`

Source listing: Appendix A.1.1

Let's start with a very basic "Hello World" example, just to go through the mechanics of building and running a small BSV program.

```
// ================================================================
/* Small Example Suite: Example 1a
   Very basic 'Hello World' example just to go through the
   mechanics of actually building and running a small BSV program
   (and not intended to teach anything about hardware!).
   ============================================================ */

package Tb;

(* synthesize *)
module mkTb (Empty);

   rule greet;
      $display ("Hello World!");
      $finish (0);
   endrule

endmodule: mkTb

endpackage: Tb
```

2.1.1   Comments

Single-line comments start with two slashes (//) and go to the end of the line. Multi-line comments can be written with the slash-star opening bracket and the star-slash closing bracket (/* `comment here` */).

2.1.2   Package

All your code must be organized into packages, which are like *namespaces*. The BSV compiler and other tools assume that there is one package per file, and they use the package name to derive the file name. The file name is `<package name>.bsv`, which in this case would be `Tb.bsv` ("Tb" for "Testbench"). Package structure does not directly correlate with hardware structure. The hardware structure is specified by the module hierarchy and constructs.

2.1.3   (* synthesize *)

Items contained within (* and *) are BSV *attributes*, used to guide the compiler in its decisions. The (*synthesize*) attribute specifies that the module that follows is synthesized separately into a hardware (Verilog) module. The top-level module of a design must be a synthesis boundary. Interior modules can also represent synthesis boundaries. The design in this example has only one module.

You can also use the -g command-line flag when compiling to specify modules that are synthesis boundaries, but we recommend using the (* synthesize *) attribute in the source text instead.

2.1.4   Module definition

The statement:

```
    module mkTb (Empty);
```

defines a module named `mkTb` providing an `Empty` interface. By convention, we call our modules "mkFoo", for "make Foo", to suggest that the module can be instantiated multiple times, possibly with different parameters. `Empty` is a pre-defined interface with no methods. The keyword `Empty` is not required, the statement could also be written as:

```
    module mkTb();
```

BSV does not have input, output, and inout pins like Verilog and VHDL. Signals and buses are driven in and out of modules with methods; these methods are grouped together into interfaces. An `Empty` interface has no inputs or outputs.

Modules and interfaces form the heart of BSV. Modules and interfaces turn into actual hardware. A module consists of three things: state, rules that operate on that state, and an interface to the outside world (surrounding hierarchy). A module definition specifies a scheme that can be instantiated multiple times.

2.1.5 Rule

The statement:

```
rule greet;
   $display ("Hello World!");
   $finish (0);
endrule
```

defines a rule named `greet` with no explicit rule condition. (An explicit rule condition is a boolean expression in parentheses after the rule name; when omitted, it defaults to `True`.)

Rules are used to describe how data is moved from one state to another, instead of the Verilog method of using `always` blocks. Rules have two components:

- **Rule conditions:** Boolean expressions which determine when the rule is enabled.

- **Rule body:** a set of actions which describe state transitions

In this simple example, the name of the rule is `greet` and there are no explicit conditions on the rule; the rule is always enabled. There are also no state transitions in the rule, just invocations of the system functions `$display` and `$finish`.

A primary feature of rules in BSV is that they are atomic; each enabled rule can be considered individually to understand how it maintains or transforms state. Rules are indivisible and a state transition is described completely by the actions within the rule. Atomicity allows the functional correctness of a design to be determined by looking at each of the rules in isolation, without considering the actions of other rules. This one-rule-at-a-time semantics greatly simplifies the process of determining the functional correctness of a design. In the hardware implementation compiled by the BSV compiler, multiple rules will execute concurrently. The compiler ensures the actual behavior is consistent with the logical behavior, thus preserving functional correctness while achieving performance goals. Rules and rule scheduling are discussed in more detail in Section 5.2.

Each rule is terminated with an `endrule`, which can optionally include the name of the rule. We could have written the `endrule` as:

```
endrule: greet
```

2.1.6 System tasks

BSV supports a number of Verilog's system tasks and functions. In this example we have two system tasks:

- `$display` invokes the system task to display a literal String (like "printf()" in C, it does formatted output during simulation);

- `$finish(0)` finishes the simulation with an exit value of 0 from the simulation process (like "exit()" in C, it terminates the simulation).

The module and package names also require "end" brackets, and as with the rule statement, the repetition of the module and package names are optional.

2.2   Building the design

2.2.1   Bluespec Development Workstation (BDW)

The Bluespec Development Workstation (BDW) is an integrated graphical design environment for creating, building, analyzing, and simulating BSV designs. The development workstation includes all Bluespec tools and accesses third-party tools such as simulators, waveform viewers, and text editors.

As we saw above, a BSV program consists of one or more outermost constructs called packages; all BSV code must be inside packages and there is a one package per file. When using the workstation you will have an additional file, a project file (*projectname*.`bspec`), which is a saved collection of options and parameters. Only the workstation uses the project file; if you use BSV completely from the Unix command line you need not have a project file.

In this guide we will assume you are using the Bluespec Development Workstation. For more details on the workstation and complete commands and instructions on using the tools from the command line, please refer to the *Bluespec SystemVerilog User Guide*.

2.2.2   Components of a BSV design

The following file types are generated when building a design with BSV. Not all file types will be generated for all designs; some depend on the target simulation environment (Bluesim or Verilog). Bluesim is a cycle simulator for BSV designs, included in your BSV release.

File Types in a BSV Design			
File Type	Description	Bluesim	Verilog
`.bsv`	BSV source File	√	√
`.bspec`	Bluespec Development Workstation project File	√	√
`.bi,` `.bo,` `.ba`	Intermediate files not directly viewed by the user	√	√
`.v`	Generated Verilog file		√
The `.h`, and `.cxx` files are intermediate files not directly viewed by the user			
`.h`	C++ header files	√	
`.cxx`	Generated C++ source file	√	
`.o`	Compiled object files	√	
`.so`	Compiled shared object files	√	

Note that the workstation and the compiler have various flags to direct the intermediate files into various sub-directories to keep your project organized and uncluttered.

2.2.3 Overview of the BSV build process

Figure 2.1 illustrates the following steps in building a BSV design:

1. A designer writes a BSV program. It may optionally include Verilog, SystemVerilog, VHDL, and C components.

2. The BSV program is compiled into a Verilog or Bluesim specification. This step has two distinct stages:

   (a) pre-elaboration - parsing and type checking
   (b) post-elaboration - code generation

3. The compilation output is either linked into a simulation environment or processed by a synthesis tool.

Once you've generated the Verilog or Bluesim implementation, the workstation provides the following tools to help analyze your design:

- Interface with an external waveform viewer with additional Bluespec-provided annotations, including structure and type definitions

- Schedule Analysis viewer providing multiple perspectives of a module's schedule

- Scheduling graphs displaying schedules, conflicts, and dependencies among rules and methods.

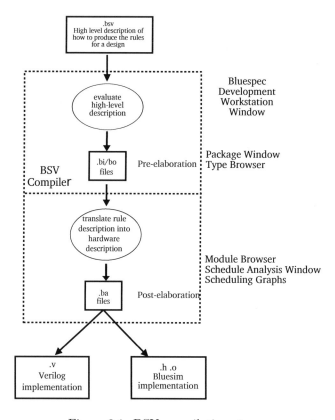

Figure 2.1: BSV compilation stages

2.2.4   Create a project file

From a command prompt, execute the command `bluespec` to start the Bluespec Development Workstation:

`bluespec`

The main window, shown in Figure 2.2, is the control center of the workstation. From this window you can manage projects, set project options, and monitor status while working in the workstation. The window displays all commands executed in addition to all warnings, errors, and messages generated by the BSV compiler and the workstation.

We're going to create a new project for our example. Select **New** from the **Project** pull-down menu. In the **New Project** dialog window select the directory containing the example file `Tb.bsv` and enter a project name. Let's call it `Example`. The project file will then be named `Example.bspec`.

After you press `Save`, the **Project Options** window will open so you can set the options and parameters for the project. On the **Files** tab, enter the **Top file** (`Tb.bsv`) and **Top module** (`mkTb`).

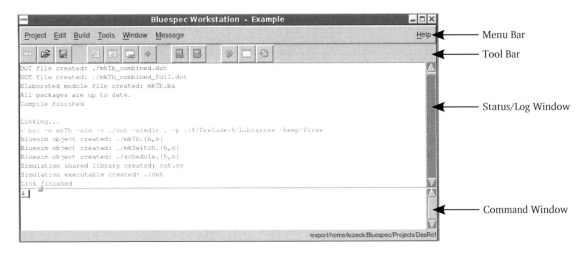

Figure 2.2: Main window components

Browse through the remaining tabs to familiarize yourself with the different options and note where they are entered.

2.2.5   Compile

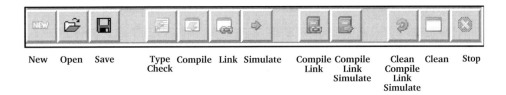

Figure 2.3: Workstation Toolbar

Building a project includes running the compiler and simulating designs. All build options are available from the **Build** menu and the toolbar, shown in Figure 2.3.

Some options combine multiple build actions into single option or button, for example, **Compile +
Link**.

Compile the design by selecting **Compile** from the **Build** menu, or by using the **Compile** button on the toolbar and make sure you have no errors in your file.

To simulate the design, run the compile, link, and simulate steps. You will see the "Hello World!" message in the workstation message window.

In this manual: Source code files (`.bsv`) and Bluespec Development Workstation project files (`.bspec`) are provided for all examples in this manual. Open the project (`.bspec`) file to run any of the examples. To run the provided `Hello World` example from workstation, open the file `hello_world.bspec` from the directory `getting_started/hello_world`. The source code and directory location for all examples are in the Appendix A.

2.3   Multiple modules in a single package

Source directory: `getting_started/mult_modules`

Source listing: Appendix A.1.2

This next example looks at a design comprised of two modules, a small testbench module communicating with a design under test (DUT) module through its interface. To use the example files provided, open the workstation project file `getting_started/mult_modules/mult_modules.bspec`. You'll notice that since both modules are still in the same package and the name of the top module is still `mkTb`, the `.bspec` file looks the same as in our previous example.

This example defines two modules:

- `mkModuleDeepThought` which receives three input values in the interface, `x, y, z`, calculates and returns, through the method `the_answer`, the sum of the values.

- `mkTb` the testbench which invokes a method in the interface on some input values and displays the returned value.

The modules communicate through the interface `Ifc_type`.

```
package Tb;

(* synthesize *)
module mkTb (Empty);

   Ifc_type  ifc <- mkModuleDeepThought;

   rule theUltimateAnswer;
    $display ("Hello World! The answer is: %0d",ifc.the_answer (10, 15, 17));
    $finish (0);
   endrule

endmodule: mkTb

interface Ifc_type;
   method int the_answer (int x, int y, int z);
endinterface: Ifc_type

(* synthesize *)
module mkModuleDeepThought (Ifc_type);

   method int the_answer (int x, int y, int z);
      return x + y + z;
   endmethod

endmodule: mkModuleDeepThought
endpackage: Tb
```

2.3.1 Interfaces

Interfaces provide a means to group wires into bundles with specified uses, described by methods. BSV does not have input, output, and inout pins like Verilog and VHDL. Signals and buses are driven in and out of modules with methods. Most modules have several methods to drive all the buses in and out of a module. These methods are grouped together into interfaces.

In SystemVerilog and BSV we separate the declaration of an interface from the definition of a module. This allows us to define common interfaces that can be offered by multiple modules, without requiring the repetition of the declaration in each of the implementation modules. All modules which share a common interface have the same methods, and therefore the same number and types of inputs and outputs.

It is important to distinguish between interface declarations and interface instances. A design may contain one or more interface declarations, and each of these may have multiple instances. An interface declaration can be thought of as declaring a new type similar to a struct type, where the members are all method prototypes. The declaration states the new type, while the definition of how the methods are implemented is contained within the module that contains the interface instance.

A method can be thought of like a function, i.e., it is a procedure that takes zero or more arguments and returns a result. Thus, method declarations inside interface declarations look a lot like function prototypes, the only difference being the use of the keyword method instead of the keyword function. Each method represents one kind of transaction between a module and its clients. When translated into RTL, each method becomes a bundle of wires. The definition of how a method is implemented is contained in the interface definition within the module and may be different in each module where the common interface is used. However the wires making up the interface, that is the inputs, outputs and inouts, and the methods defined, will be the same in all instances.

Let's look at the definition of interface Ifc_type. Interface names always start with an upper-case letter, because they are a data type, in this case a user-defined data type. The interface declares a value method named the_answer with three int arguments returning an int result. How the result is calculated is left to the method definition.

```
interface Ifc_type;
    method int the_answer (int x, int y, int z);
endinterface: Ifc_type
```

The module mkModuleDeepThought provides an interface of type Ifc_type, as shown in the module definition statement:

```
module mkModuleDeepThought (Ifc_type);
```

To instantiate a module and obtain a handle on its interface, use a statement of the form:

Interface_type identifier <- *module_name* ;

In the testbench module (mkTb), an instance of the interface type Ifc_type named ifc is created by instantiating a module which provides this interface:

```
    Ifc_type  ifc <- mkModuleDeepThought;
```

The fundamental difference between a method and a function is that a method also carries with it *implicit conditions*, the handshaking signals and logic which are automatically generated by the BSV compiler.

BSV classifies interface methods into three types:

- **Value Methods:** These are methods which return a value to the caller, and have no "actions" (i.e., when these methods are called, there is no change of state, no side-effect).

- **Action Methods:** These are methods which cause actions (state changes) to occur. One may consider these as input methods, since they typically take data into the module.

- **ActionValue Methods:** These methods couple Action and Value methods, causing an action (state change) to occur *and* they return a value to the caller.

The value method `the_answer` takes the integer arguments `x`, `y`, `z` and calculates the sum of the arguments. Since it is a value method, we know that it does not change any state elements. The definition of the method, that is the calculation of the return value based on the arguments, is defined in the module `mkModuleDeepThought`.

```
method int the_answer (int x, int y, int z);
   return x + y + z;
endmethod
```

2.4 Multiple packages in a single design

Source directory: `getting_started/mult_packages`

Source listing:Appendix A.1.3

This example separates the two module definitions in our previous example into separate packages: `Tb.bsv` and `DeepThought.bsv`.

The only change required is add an `import` statement to the package Tb to import all declarations from the package DeepThought:

```
import DeepThought::*;
```

The top file in the design is still `Tb.bsv` and the top module is still `mkTb`. To modify the current example, you can create the new file `DeepThought.bsv` in the same directory as the file `Tb.bsv`. From the **Project Files** window you can select **File→New** to create a new file from within the workstation.

To run the example provided, open the project file:

`getting_started/mult_modules/mult_modules.bspec`

From the workstation you can review the source code, compile, and simulate the example.

Chapter 3

Data types

3.1 Type

Data types are an abstraction. Ultimately, all computation, whether in hardware or software, is done on bits, but it is preferable to think in terms of the concepts like integers, floating point numbers, fixed point numbers, Ethernet packets, video frames, and the like. Strong type-checking prevents unintentional "misinterpretation" of bits, such as taking the square root of an Ethernet packet, or the IP address of a vector of floating-point numbers.

A second and important reason for this abstraction is *representation independence*. For example, the same abstract `struct` type can have many possible representations in bits (ordering of fields, little- vs. big-endian, padding for byte and word alignment, and so on). Different representations may be more appropriate depending on the application environment. By separating out the type abstraction from its bit representation, we can easily change representations without modifying or breaking the rest of the program.

Types play a central role in BSV. Every variable and every expression has a type and variables can only be assigned values which have compatible types. The BSV compiler provides a strong, static type-checking environment. Type checking, which occurs before program elaboration or execution, ensures that object types are compatible and that conversion functions are valid for the context.

Data types in BSV are case sensitive. The first character of a type is always *uppercase* (there are only two exceptions: the types `int` and `bit`, compatibility with Verilog). The first character of a variable name is always *lowercase*. A common source of errors for beginners in BSV is to use an uppercase name where BSV expects a lower case, and vice versa.

3.2 Uppercase and lowercase in type and value identifiers

In BSV, the case of the first letter of an identifier is significant, whether used to describe types or values. The general principle is that an uppercase first letter introduces a *constant*, whereas a lowercase first letter introduces a *variable*.

A constant type is a particular type, such as `Bool`, `Int`, `UInt`, `Server`, `Action`, `Module`, `Rule`, and so on. A type variable is used to represent a *polymorphic* type, similar to "generic" or "template" types in other languages (see Section 9 for more on polymorphism).

A constant value is a particular symbolic value, such as `True`, `False`, the symbolic labels declared in all `enum` types, the constructors associated with any `struct` or `tagged union` type, and so on. A type variable is the familiar identifier bound to some particular value, such as `x`, `foo`, `mkTb`, `mkDut`, and so on.

3.3 Typeclasses and overloading

[This section can be skimmed lightly on first reading. Overloading is used in a fairly lightweight manner in subsequent sections, for which this section provides some intuition and context.]

BSV makes very effective use of systematic, user-extensible *overloading*. For example, instead of having *ad hoc* definitions of how various data types are represented in bits, in BSV each such data type has a pair of overloaded functions called `pack` and `unpack` that convert from the abstract view of the type into bits and vice versa. Further, because overloading is user-extensible, the user can specify the definitions of these functions precisely, and thus specify representations precisely.

In BSV, we say that there is a "typeclass" called `Bits#(`t,n`)`, which can be regarded as a set of types. Any type t in this typeclass has definitions for bit representation in n bits, i.e., it has definitions for the overloaded functions:

```
function  Bit#(n) pack   (t x);
function  t       unpack (Bit#(n) b);
```

Each member type t in a typeclass is called an *instance* of the typeclass. The functions `pack` and `unpack` are called *overloaded* because the same function names are used for different types, and the compiler figures out the appropriate ones to use based on the actual types of their arguments and results.

In most cases in the examples below, we will use a shortcut to define bit representations. Instead of a full-blown `instance` declaration, you will often see the phrase "`deriving (Bits)`" appended to a type definition, indicating to the compiler to pick the obvious, canonical bit representation. Thus, we only write a full-blown instance declaration for `Bits` if we need a non-standard representation.

Thus, a typeclass is a construct which implements overloading across related data types. Overloading is the ability to use a common function name or operator on a collection of types, with the specific function or operator being selected by the compiler based on the types on which it is actually used (this process is called "overloading resolution"). A typeclass may include multiple data types; all data types within the typeclass share functions and operators, hence the function names within a typeclass are *overloaded* across the various typeclass members.

Another example: the `Ord` typeclass defines the class of types for which an *order* is defined, allowing comparison operations. The `<` function is meaningful on types that can be compared such as bits, numeric types, or vectors. A complete definition of an instance of `Ord` requires defining how items in the class are compared, either through defining the operator `<=` or the function `compare`. The implementation of how you compare the various types may be different, but all the members of the

Ord typeclass define and allow comparison. For example, if you had a MemRequest data type, you could make it an instance of the Ord typeclass by defining the meaning of <= (perhaps defined to mean a request priority).

A type can be declared to be an *instance* of a type class. The instance declaration must specify the implementation of the overloaded identifiers and functions of the type class.

[C++ gurus: a BSV typeclass is like a "virtual class" in C++. The overloaded identifiers are the "virtual members" of the class. A BSV typeclass instance is like a class inheriting from the virtual class, and providing the definitions of the virtual members.]

3.4 Pre-defined type classes

BSV defines a common set of type classes. A data type may belong to zero or many type classes. Users can also define their own type classes. The following type classes are pre-defined in BSV:

Provided Type Classes	
Bits	Types that can be converted to bit vectors and back.
Eq	Types on which equality is defined.
Literal	Types which can be created from integer literals.
RealLiteral	Types which can be created from real literals.
Arith	Types on which arithmetic operations are defined.
Ord	Types on which comparison operations are defined.
Bounded	Types with a finite range.
Bitwise	Types on which bitwise operations are defined.
BitReduction	Types on which bitwise operations on a single operand to produce a single bit result are defined.
BitExtend	Types on which extend operations are defined.

3.5 Data type conversion functions

To convert across types in BSV, the following overloaded functions will often suffice. During type checking, the compiler resolves these functions to a particular type instance. Note that a design containing excessive type conversions usually indicates a poor choice of types in the design. The conversion utilities presented here do not incur logic overhead.

The most common overloaded functions are pack and unpack, used to convert to and from the type Bit. For bit-extensions, We recommend you use extend instead of either the zeroExtend or signExtend functions, as it is overloaded and will resolve to the appropriate extension function for you automatically.

Data Type Conversion Functions		
Function	Type Class	Description
pack	Bits	Converts (packs) *from* types in the Bits class, including Bool, Int, and UInt *to* Bit.
unpack	Bits	Converts (unpacks) *from* Bit *to* other types in Bits class, including Bool, Int and UInt.
extend	BitExtend	Performs either a zeroExtend or a signExtend as appropriate, depending on the data type of the argument.
zeroExtend	BitExtend	Increases the bit width of an expression by padding the most significant end with zeros. Normally used for unsigned values.
signExtend	BitExtend	Increases the bit width of an expression by replicating the most significant bit. Normally used for signed values.
truncate	BitExtend	Shortens the bit width to a particular width.
fromInteger	Literal	Converts from an Integer type to any type defined in the Literal type class. Integers are most often used during static elaboration.
fromReal	RealLiteral	Converts from a Real type to any type defined in the RealLiteral type class.
valueOf	Any numeric type	Converts from a numeric type to an Integer. Numeric types are the n used in types such as Bit#(n) and Vector#(n, t).

3.6   Some common scalar types

The following pre-defined scalar types are all in the Bits typeclass. All hardware types (values carried on wires, stored in registers, FIFOs or memories, etc.) must be in the Bits typeclass (but note, arbitrary types can be used during static elaboration).

3.6.1   Bit#(n)

The most basic type is Bit#(n), a polymorphic data type defining a type consisting of n bits. Verilog programmers will naturally want to only ever use Bit#(n) as this gives them the complete bit selecting and operating freedom they are used to with Verilog. But Verilog also allows a lot of dubious operations on data, for example, automatic truncation. And more importantly, a Bit#(n) doesn't really indicate anything about the semantics of the data.

In BSV, we want to take advantage of strong typing by using meaningful data types. This decreases bugs and helps find bugs earlier in the design cycle, at compile time, rather than at debug or tapeout time.

The familiar Verilog type bit is a synonym for Bit#(1).

3.6.2 Bool

We use `Bool` as opposed to `bit` or `Bit#(1)` to define true/false logical operations. In BSV, all "conditions" (in rules, methods, if statements, while statements, etc.) require Bool types.

```
typedef enum {False, True} Bool;
```

The implementation of a Bool is one bit. But, be aware, Bool is *not* considered type-equivalent to `Bit#(1)`. Use `Bool` for `True` or `False` values and `Bit#(1)` for variables with a value of 0 or 1. You cannot automatically convert from a `Bool` to a `Bit#(1)`. Bit-wise operations cannot be performed on a `Bool` type.

The following functions are defined for `Bool` types and return a value of `True` or `False`.

- `not` or `!`: Returns `True` if x is false, returns `False` if x is true
- `&&`: Returns `True` if x *and* y are true, else returns `False`
- `||`: Returns `True` if x *or* y are true, else returns `False`

3.6.3 UInt#(n)

`UInt#(n)` is an unsigned integer of defined bit length. So a `UInt#(16)` includes values 0 to `'hFFFF` unsigned. It can never be negative or less than zero. This is a common type to use for index counters, etc. where you are counting things that will never be negative.

`UInt#(n)` has the logical bitwise operators defined, as well as `zeroExtend` to append zero bits to the MSB, and `truncate` to remove bits from the MSB.

3.6.4 Int#(n)

`Int#(n)` is a signed integer of defined bit length used to represent and compare variables which are interpreted as signed quantities. The familiar Verilog/C type `int` is a synonym for `Int#(32)`.

Use `signExtend` to create larger integers by appending bits to the MSB by replicating the sign bit. Use `truncate` to remove bits from the MSB.

3.7 Using abstract types instead of Bit

Source directory: `data/bit_types`

Source listing: Appendix A.2.1

This example looks at using the abstract types `Bool`, `UInt#(n)`, and `Int#(n)` instead of `Bit#(n)`. Each rule examines a different type. The complete example, found in Appendix A.2.1, contains multiple display statements to help you understand what is happening in the source code when you simulate the design.

```
package Tb;

(* synthesize *)
module mkTb (Empty);

    Reg#(int) step <- mkReg(0);

    Reg#(Int#(16))  int16  <- mkReg('h800);
    Reg#(UInt#(16)) uint16 <- mkReg('h800);
```

Rule step0 examines the UInt data type.

```
    rule step0 ( step == 0 );
        //  UInt#(16) foo = -1;      //invalid value to initialize
        UInt#(16) foo = 'h1fff;  //valid value to initialize
```

Since a UInt#(n) is an unsigned int, you cannot initialize it with a negative number. The first line
in rule step0 is commented out in the code; uncomment it and you will get the error message:

```
Error: "Tb.bsv", line 40, column 19: (T0051)
   Literal '-1' is not a valid UInt#(16).
```

UInt#(n) does not allow access to individual bits, as such [ ] is not a legal construct for UInt#(n)
as it is for Bit#(n).

```
        foo = foo & 5;
        $display("foo[0] = %x", foo[0]); //invalid statement

        foo = 'hffff;
        $display("foo = %x", foo);       //valid statement

        // and it wraps just like you would expect it to
        foo = foo + 1;
        $display("foo = %x", foo);
```

Because this is an unsigned number, the value can never be less than zero. This display:

```
        $display("fooneg = %x", foo < 0);
```

is displayed the same as:

```
        $display("fooneg = %x", 1'd0 ) ;
```

One way to get min and max values from UInts is to use unpack:

```
        UInt#(16) maxUInt16 = unpack('1);   // all ones
        UInt#(16) minUInt16 = unpack(0);
```

The next rule, step1, examines the Int#(n) data type. Int#(n) has similar restrictions as UInt#(n), but it does include negative numbers. Again, the quick way to get min and max values is use unpack:

```
        Int#(16) maxInt16 = unpack({1'b0,'1});   // 'h011111
        Int#(16) minInt16 = unpack({1'b1,'0});   // 'h100000
```

As with UInt#(n), you cannot perform bit manipulation directly. The following statement will give an error:

```
        $display("error1 = %x", minInt16[4]);
```

Because the type Int#(n) (as well as the type UInt#(n) is in the Arith type class, you can perform the operations +, -, /, *, %. But be careful when using multiplication and division functions as they may create long timing paths.

```
        $display("maxInt16/4 = %x", maxInt16 / 4);
        int16 <= int16 / 4;
```

Rule step2 examines the exceptions to the rule that data types are always capitalized: int and bit. int is a special case defined for Int#(32) while bit is predefined as Bit#(1).

```
        Int#(32) bar = 10;
        int foo = bar;

        bit onebit = 1;
        Bit#(1) anotherbit = onebit;
    endrule
```

The final rule in this example, step3, examines the use of Bool.

```
       Bool b1 = True;
       Bool b2 = True;
       Bool b3 = b1 && b2;
```

Remember, the conditional in an if statements needs to be of type Bool, not bit.

```
    // if (onebit)                 // uncomment this to see the type error
    //    $display("onebit is a 1");

    if ( b1 )
       $display("b1 is True");
```

If you are testing a bit value, you have to turn that bit into an expression returning a Bool type in order to test:

```
    bit onebit = 1;
    if (onebit == 1)
       $display("onebit is a 1");
```

3.8   Integers

Source directory: data/integers

Source listing: Appendix A.2.2

The Integer type represents mathematical integers. As in mathematics, they are unbounded (in practice, they are bounded only by your computer's memory size and your own patience!). Integer is not part of the Bits typeclass and hence cannot be used for hardware values. However, it is frequently used in BSV programs as part of static elaboration for loop iterators and array indexes; all values are resolved at compile time. Integer is different from UInt#(n) and Int#(n), both of which are bounded, belong to the Bits typeclass and are synthesizable into hardware representations.

For example, we can create an array of 4 elements, and use the Integer value to index the array statically:

```
Integer inx = 0;

Bool arr1[4];
arr1[inx] = True;

inx = inx + 1;
arr1[inx] = False;

inx = inx + 1;
arr1[inx] = True;

inx = inx + 1;
arr1[inx] = True;
```

You can also create a loop with the integer variable. In order to convert an `Integer` value to a sized value, such as `Int#(32)`, you need to use the `fromInteger` overloaded conversion function.

```
Int#(32) arr2[16];
for (Integer i=0; i<16; i=i+1)
    arr2[i] = fromInteger(i);
```

An `Integer` value can be used in mathematical operations:

```
Integer foo = 10;
foo = foo + 1;
foo = foo * 5;
Bit#(16) var1 = fromInteger( foo );
```

`Integer` is in the `Arith` typeclass, so all arithmetic functions defined for the typeclass are defined for `Integer` values. The following additional mathematical functions are also defined for `Integer`:

- `div`: Element x is divided by element y and the result is rounded toward negative infinity. Division by 0 is undefined.

- `mod`: Element x is divided by element y using the `div` function and the remainder is returned as an `Integer` value. `div` and `mod` satisfy the identity $div(x, y) * y + mod(x, y) == x$. Division by 0 is undefined.

- `quot`: Element x is divided by element y and the result is truncated (rounded towards 0). Division by 0 is undefined.

- `rem`: Element x is divided by element y using the `quot` function and the remainder is returned as an `Integer` value. `quot` and `rem` satisfy the identity $quot(x, y) * y + rem(x, y) == x$. Division by 0 is undefined.

- exp The element `base` is raised to the `pwr` power and the exponential value `exp` is returned as type `Integer`.

3.9   Strings

Source directory: `data/strings`

Source listing: Appendix A.2.3

BSV provides support for Strings. Remember that, by definition, the compiler is creating synthesizable logic. So Strings are just ASCII characters joined into a list. They must have a fixed size. String literals are enclosed into double quotes and must be contained on a single line of source code.

Special characters may be inserted in string literals, with the following backslash escape sequences:

\n	newline
\t	tab
\\	backslash
\"	double quote
\v	vertical tab
\f	form feed
\a	bell
\OOO	exactly 3 octal digits (8-bit character code)
\x$HH$	exactly 2 hexadecimal digits (8-bit character code)

We can define an Action function to pass a string to a function. This example shows setting a string to a value and passing a string to a function. We can use + to concantenate strings.

```
function Action passString( String s );
   return action
           $display("passString => ", s);
           endaction;
endfunction

rule init ( step == 0 );
   ...
   passString( "String passed to a function" );
endrule
```

Chapter 4

Variables, assignments, and combinational circuits

4.1 Variable declaration and initialization

Source directory: `variable/declaration`

Source listing: Appendix A.3.1

BSV defines two types of initializations:

- *pure* or *value* initializations using (=): the value on the right-hand side is bound to the variable on the left-hand side.

- *side-effecting* initializations using (<-): the right-hand side is an expression that has a side effect and also returns a value. The returned value is bound to the variable on the left-hand side.

  There are two common kinds of side effects:

  - instantiating a module and returning its interface
  - invoking an ActionValue and returning its value.

To demonstrate variable declaration and initialization syntax, we'll be working with the example in Appendix A.3.1. The syntax in this example is more important than the functionality.

Our example has a structure representing a 2-dimensional coordinate:

```
typedef struct { int x; int y; } Coord
    deriving (Bits);
```

We'll describe structs in more detail in Section 10.1.3, but will rely here on their familiarity from C. The above defines a new type `Coord` with two `int` fields (members) called `x` and `y`. The "deriving (Bits)" phrase instructs the compiler to pick a canonical representation, here the obvious 64-bit value concatenating the two 32-bit fields.

4.1.1 Side-effect initialization

One type of side-effect initialization is instantiating a module and returning its interface. Consider the following statement:

```
FIFO#(Coord) fi <- mkFIFO;
```

On the left-hand side we are specifying an interface type `FIFO#(Coord)` and a variable `fi` of that type. On the right-hand side we invoke a module instantiation `mkFIFO`. The `FIFO` interface returned by the module instantiation is bound to the variable `fi`.

When a module instantiation, such as the expression `mkFIFO` is used on the right-hand side of an `<-` statement, it instantiates the module and returns a value of the interface type, in this case `FIFO#(Coord)`. This represents a fifo containing elements of type `Coord`.

The reason for thinking of such an invocation of `mkFIFO` as a side-effect is that each that each time it is invoked in this way, it creates a new FIFO instance different from the earlier instantiations.

4.1.2 Value initialization

A value initialization assigns the value on the right-hand side to a variable on the left-hand side. It cannot be used to update a register or other state element. Consider the statement:

```
Coord delta1 = Coord { x: 10, y: 20 };
```

On the left-hand side we specify a variable named `delta1` of type `Coord`. Note that the variable name is lower case, the type name is uppercase. The right-hand side is a so-called *struct expression* (see Section 10.1.3) that evaluates to a struct value of type `Coord`. This value is assigned to the variable named `delta1`.

We could also use the `let` shorthand, leaving it up to the compiler to infer the type of `delta2`, which must be the same type as the expression on the right-hand side, `Coord`.

```
let delta2 = Coord { x: 5, y: 8 };
```

This shorthand is especially useful when the type is more complicated, but should be used with care since it reduces readability. In this example the right-hand side is a `Coord` struct expression; the type is obvious to the reader.

4.1.3 Value initialization from a `Value` method

Remember from above, the variable `fi` is of the type `FIFO#(Coord)`. Consider the following statement:

```
Coord c = fi.first();
```

On the left-hand side we specify a type `Coord` and a variable `c` of that type. The right-hand side invokes the value method `fi.first()` whose result is of type `Coord` and is assigned to c. Note that `fi.first()` is a value method, not an ActionValue method. Therefore it has no side-effect, which is why it is used in an `=` statement, and not a `<-` statement. There is no change of state with a value method.

This line could have been written using the `let` statement and the compiler would have automatically set the type of c to `Coord`:

```
let c = fi.first();
```

4.1.4 Initializing sub-interface components

The `mkTransformer` module in our example provides a `Server#(Coord, Coord)` interface.

```
module mkTransformer (Server#(Coord, Coord));
```

As defined in the package `ClientServer`[1], a `Server#()` interface has two subinterfaces, `request` and `response`:

```
interface Server#(type req_type, type resp_type);
    interface Put#(req_type)  request;
    interface Get#(resp_type) response;
endinterface: Server
```

There is a special syntax for initializing sub-interface components. `request` is the name of a sub-interface of the overall `Server#()` type for the module. This sub-interface is of the type `Put#()`. The statement:

```
interface request = toPut (fi);
```

has the same type (`Put#()`) on both sides of the `=`. There is no change of state or instantiation of a module, so the operator used is `=`.

---

[1] The `ClientServer` package is described in more detail in Section 6.8

4.1.5   Additional examples

The statement:

```
Reg#(int) cycle <- mkReg (0);
```

specifies a type `Reg#(int)` and a variable `cycle` of that type, and we initialize it by invoking the side-effect of instantiating the module using `mkReg(0)` and binding the resulting interface to the variable `cycle`. We use the <- assignment operator for instantiating modules. We'll discuss registers in more detail in Section 5.1.

The statement:

```
Server#(Coord, Coord) s <- mkTransformer;
```

specifies a type `Server#(Coord, Coord)` and a variable `s` of that type, and we initialize it by invoking the side-effect of instantiating the module using `mkTransformer` and binding the resulting interface to the variable `s`.

In the statement:

```
let c = Coord { x: rx,  y: ry };
```

we use the `let` shorthand, leaving it up to the compiler to infer that `c` has the type `Coord`, based on the expression on the right-hand side.

In the statement:

```
let c <- s.response.get ();
```

we've used `let` in a <- statement. The right-hand side expression `s.response.get()` has the type `ActionValue#(Coord)`. By using it in a <- statement we invoke the `ActionValue`'s side effect, and the returned value of type `Coord` is bound to the variable `c`. By using `let`, we allow the compiler to infer the type of `c`, which is `Coord`.

To illustrate the difference between "an operation with a side-effect" and "*invoking* an operation with a side-effect", the following statement:

```
Server#(Coord, Coord) s <- mkTransformer;
```

could also have been written as:

```
Module #(Server#(Coord, Coord)) m = mkTransformer;
Server#(Coord, Coord)                s <- m;
```

In the first line, we are not invoking the operation `mkTransformer`; we are simply binding *the operation itself* to the variable m. Observe the type of m carefully, because it reflects this. It's type is `Module #(Server #(Coord, Coord))`, not `Server #(Coord, Coord)`. The second line, by using `<-`, actually invokes the operation bound to m, which executes the side-effect (instantiating the module), and returns the interface, which is bound to s, whose type is `Server #(Coord, Coord)`.

Similarly, the following:

```
Coord c <- s.response.get ();
```

could also be written as:

```
ActionValue #(Coord) g = s.response.get;
Coord c                 <- g ();
```

illustrating the difference between assigning an `ActionValue` versus invoking an `ActionValue` and assigning its result.

The C/C++ analogy would be the difference between merely assigning a function pointer versus actually invoking the function referenced by a function pointer (by appending "()" after the pointer) and assigning the result of the function call. This begs the question: If C/C++ does it just using function-call notation and the same assignment symbol, why does BSV need two separate assignment symbols for this? The answer is that C/C++ does not make any attempt to distinguish between functions that have side-effects and functions that don't (that are "pure"). This may be fine in C/C++'s context of sequential von Neumann computing, but it is the root of much evil in parallel computing, and in hardware design we have nothing if not an extreme case of parallel computing. BSV's type system and notation is carefully designed to clearly identify and contain side-effects into well-defined contexts, and is the source of its robustness as we scale to larger and larger hardware designs.

4.2   Combinational circuits

Source directory: `variable/combcircuits`

Source listing: Appendix A.3.2

In BSV, expressions representing hardware values are really just combinational circuits. Expressions are built from constants, variables, operators and function applications. A variable is just a name for a value. Unlike languages like C/C++/C#/Java, in BSV a variable does not represent a chunk of storage in a computer's memory. In hardware terms, this is equivalent to saying that a sequence of variable assignments is an incremental way to specify a combinational circuit.

In a combinational circuit, values are carried on wires. One way of thinking about this is to consider the variable as a name for the wire. In BSV, variables are never storage elements or containers.

Every variable and expression in BSV has a type. The BSV compiler checks that constructs in the language are applied correctly according to types.

Let's look at the variable assignment semantics in the following example:

```
module mkTb (Empty);

   Reg#(int) cycle <- mkReg (0);

   rule count_cycles;
      cycle <= cycle + 1;
      if (cycle > 7) $finish(0);
   endrule

   int x = 10;

   rule r;
      int a = x;
      a = a * a;
      a = a - 5;

      if (pack(cycle)[0] == 0) a = a + 1;
      else                     a = a + 2;

      if (pack(cycle)[1:0] == 3) a = a + 3;

      for (int k = 20; k < 24; k = k + 1)
         a = a + k;

      $display ("%0d: rule r, a = %0d", cycle, a);
   endrule
endmodule: mkTb
```

Variables, like x and a in this example, never represent storage, and assignment statements with = never represent updating of storage. Variables are simply names for the values (output of combinational circuits) represented by the expressions on the right-hand sides of assignment statements. For example:

```
   a = a * a;
```

Suppose a on the right-hand side represents the output of some combinational circuit, CC. In this case, a represents the value of x, the trivial combinational circuit whose output is the constant 10.

This statement says that in the sequel, i.e. in the text following the statement within the current scope, a represents the output of a new combinational circuit, namely the output of a multiplier both of whose inputs are fed by CC. In other words, this statement incrementally describes a new combinational circuit from existing combinational circuits.

Consider the following statement in rule r:

```
if (pack(cycle)[0] == 0) a = a + 1;
else                     a = a + 2;
```

Let a before this statement represent the output of a combinational circuit CC1. This statement says that in the sequel, a will describe the output of a multiplexer whose inputs are fed from two combinational circuits, CC1+1 and CC1+2 and whose selector is fed from a combinational circuit that tests the zero'th bit of cycle for equality to zero. Consider the following statement:

```
if (pack(cycle)[1:0] == 3) a = a + 3;
```

Let a before this statement represent a combinational circuit CC2. This statement says that in the sequel, a will describe the output of a multiplexer whose inputs are fed from two combinational circuits, CC2+3 and CC2, and whose selector is fed from a combinational circuit that test the lower-order bits of cycle for equality to 3.

The loop described in the statement:

```
for (int k = 20; k < 24; k = k + 1)
    a = a + k;
```

is *unrolled* and is the same as the sequence obtained by copying the loop body:

```
a = a + 20;
a = a + 21;
a = a + 22;
a = a + 23;
```

In the sequel, a represents the output of a chain of four adders that add 20, 21, 22, and 23, respectively, to the previous combinational circuit.

If we compile and execute the complete example, found in Appendix A.3.2, we see the following output:

```
0: rule r, a = 182
1: rule r, a = 183
2: rule r, a = 182
```

```
3: rule r, a = 186
4: rule r, a = 182
5: rule r, a = 183
6: rule r, a = 182
7: rule r, a = 186
8: rule r, a = 182
```

which is consistent with our interpretation of variables and assignments.

[Compiler gurus: BSV's interpretation of variables and assignments is in essence Static Single-Assignment form (SSA), commonly used as an intermediate form in many compilers. Conceptually, repeated assignments of a variable represent a renaming of the variable and its subsequent uses to a fresh variable.]

4.2.1   Error - no type declaration

If we change the `int` declaration for `a`:

```
a = x;     // was: int a = x;
```

when we compile, we get the following error message:

```
Error: "Tb.bsv", line 33, column 17: (P0039)
    Assignment of 'a' without declaration, or non-local assignment in possibly
    wrong context (e.g. declared at evaluation time, maybe updated at runtime,
    perhaps you meant "<=" instead, etc.)
```

The compiler is complaining that we are assigning a value to `a` without any declaration. Every variable *must* first be declared, either with a type or by using `let`, before it can be used.

4.2.2   Error - no initialization

If we comment out the initialization of a:

```
int a; // was: int a = x;
a = a * a;    //line 34
```

when we compile we get the following error message:

```
Error: "Tb.bsv", line 34, column 11: (P0040)
    Use of 'a' without assignment
```

The variable on the right-hand side is now meaningless. Since `a` has not been initialized or assigned, it does not represent any combinational circuit, and so the expression `a*a` is meaningless.

Chapter 5

Rules, registers, and fifos

5.1 Defining and updating a register

Source directory: `rulesregs/register`

Source listing: Appendix A.4.1

In this section we discuss the most basic state element, the register, and how to use registers to store values. We'll also examine rules and how they are used to describe state transitions.

This example defines a register containing `int` values; the name of the register is `x`:

```
package Tb;

(* synthesize *)
module mkTb (Empty);

    Reg#(int) x <- mkReg (23);

    rule countup (x < 30);
        int y = x + 1;
        x <= x + 1;
        $display ("x = %0d, y = %0d", x, y);
    endrule

    rule done (x >= 30);
        $finish (0);
    endrule

endmodule: mkTb

endpackage: Tb
```

5.1.1   Registers

All state instances in BSV are specified explicitly using module instantiation. The most elementary module available in BSV is the register, which has a `Reg` interface. One way to instantiate a register is with the `mkReg` module, whose single parameter is the initial value of the register. The statement:

```
Reg#(int) x <- mkReg (23);
```

uses the module `mkReg` to instantiate a register (a sub-module) with an initial (reset) value of 23. The instantiation *returns* or *provides* or *yields* an interface of type `Reg#(int)` which we bind to the variable x. In other words, x is bound to an interface containing the methods `_read()` and `_write()`, which constitute a register interface. Further, this definition states that only values of type `int` can be read from or written to x.

The register interface is defined as follows:

```
interface Reg#(type a);
    method Action  _write (a x);
    method a       _read;
endinterface: Reg
```

Since registers are so common, BSV provides simplified notations for reading and writing registers. For writes, instead of using the register `_write` method explicitly, use the non-blocking assignment notation. For reads, instead of using the register `_read` method explicitly, just mention the register interface in an expression.

The non-blocking assignment (`<=`) is a shorthand notation for a `_write` method, used for dynamic assignment of a value to a register. The following example demonstrates the shorthand for both a register read and write.

```
x <= x+1;    // same as  x._write(x._read + 1);
```

5.1.2   Updating the register

Rule execution is logically instantaneous. All state read by a rule, including a register read, is logically what was in effect just before the rule. All state updated in a rule, including a register write, is logically visible only after the rule executes.

```
rule countup (x < 30);
    int y = x + 1;
    x <= x + 1;
    $display ("x = %0d, y = %0d", x, y);
endrule
```

This rule can *fire* or *execute* only when x < 30. When the example is run, the $display statement will show different values for x and y. As discussed in Section 4.2, y is just a name for the output of the combinational circuit represented by the expression x + 1, and that is what is displayed as its value. But state element updates are visible only after the rule is executed. The assignment x <= x + 1 is a shorthand for the Action x._write(x._read + 1). The effect of an Action is only visible after the rule executes. Thus, the value of x read by the $display statement is still the old value of x, before the register is written. More generally, the textual order of Actions in a rule is irrelevant—all of their effects happen simultaneously (and *atomically*) and are visible only *after* the rule's execution.

5.2 Rules and their semantics: atomicity and orderings

BSV does not have always blocks like Verilog. Instead, *rules* are used to describe all behavior (how state evolves over time). Rules are made up of two components:

- Rule Condition: a boolean expression which determines if the rule body is allowed to execute ("fire")

- Rule Body: a set of actions which describe the state updates that occur when the rule fires

As in many languages, there is a *logical view* or *semantic view* of BSV behavior which is distinct from the *implementation*. The former is how designers/programmers think about their source code. The latter is the output of the compiler (in our case, Verilog or Bluesim) and, ideally, should not be of concern to someone trying to explain or understand a BSV program.

First, we think of a rule's execution as logically *instantaneous*, *complete*, and *ordered* with respect to the execution of all other rules. By "instantaneous" we mean, conceptually, that all the actions in a rule body occur at a single, common instant—there is no sequencing of actions within a rule. By "complete" we mean that, when fired, the entire rule body executes; there is no concept of "partial" execution of a rule body. By "ordered" we mean that each rule execution conceptually occurs either before or after every other rule execution, never simultaneously.

With these properties we also say that each rule is *atomic* with respect to all other rules, or that each rule is an *atomic transaction*. Atomic transactions date back at least to the 1970s and are, to this day, by far the most effective tool known to Computer Science for describing complex concurrent processes operating on shared data. The reason is that their properties (particularly ordering) greatly simplify reasoning about the correctness of a parallel program, because one does not have to worry about potential interleavings of the actions of multiple rules. It is precisely such unwanted interleavings that cause deadly "race conditions" that are the bane of most parallel programming not based on atomic transactions (see the excellent article *The Problem with Threads* by Prof. Ed Lee in IEEE Computer 39:5, May 2006, pp. 33-42 that describes these issues in great detail).

In BSV, the logical sequence of rule executions are further grouped into sub-sequences called "clocks". There is no *a priori* limit on how many rules can fire within a clock. However, instead of allowing arbitrary sub-sequences of rules within a clock, we impose some constraints based on pragmatic hardware considerations. These constraints are:

- Each rule fires at most once within a clock.

- Certain pairs of rules, which we will call *conflicting*, cannot both fire in the same clock.

Conflicts arise due to hardware considerations. For example, in hardware each state element can change state only once per clock. Thus, we cannot have a rule sub-sequence within a clock where an one rule's state update is read by a later rule in the same clock. We call this a "rule ordering conflict". Similarly, certain hardware resource constraints, such as the fact that a hardware wire can only be driven with one value in each clock, will preclude some pairs of rules from executing together in a clock because they both need that resource. We call this a "rule resource conflict".

The BSV compiler produces hardware that respects the logical semantics and also the constraints due to conflicts. To achieve this, it synthesizes a combinational hardware *scheduler* that controls rule execution accordingly. Since rule conditions are expressions of type `Bool`, BSV's type system guarantees that they cannot have any side-effects, and so the hardware scheduler is combinational, i.e., its decisions about firing rules involve no state changes.

Note that the scheduler logic produced by the BSV compiler is not "extra" logic due to rule semantics. Logic to manage concurrent access to shared state exists in any hardware design, whether created by drawing schematics or writing Verilog/VHDL by hand. We are here only giving a vocabulary and semantic framework within which to describe such necessary logic. The user should not fear that there is some extra "overhead" due to working with rules.

One might imagine that rule ordering conflicts will prevent one rule from communicating with another rule within the same clock, since the former's state update is visible to the latter only in the next clock. However, later in Section 8 on RWires we shall see how it is possible for one rule to communicate with another within the same clock.

The basic semantics of rule atomicity (instantaneous, complete and ordered) are enough to answer most questions about functional correctness. Until this point one can consider the semantics to be "untimed". The further considerations of grouping into clocks, and conflicts, are typically needed to answer questions about performance (latency, bandwidth, and so on).

5.3   Composition of rules

Source directory: `rulesregs/rules`

Source listing: Appendix A.4.2

This example looks at when rules might conflict - that is rules that can or can't be composed in parallel. In this section we'll be discussing the following four `int` registers:

```
Reg#(int) x1    <- mkReg (10);
Reg#(int) y1    <- mkReg (100);

Reg#(int) x2    <- mkReg (10);
Reg#(int) y2    <- mkReg (100);
```

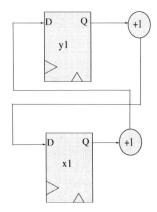

Figure 5.1: Single Rule Swapping Values

Our first example looks at a single rule swapping values:

```
rule r1;
    $display ("swap");
    x1 <= y1 + 1;
    y1 <= x1 + 1;
endrule
```

The body of this rule contains two register-write Actions. These actions occur simultaneously and instantaneously. The values of x1 and y1 are read before the rule-firing instant, and write of the the new (+1) values takes effect after the rule-firing instant. After the rule executes, x1 has 1+ the *old* value of y1, and y1 has 1+ the *old* value of x1. It's effectively a *swap* of x1 and y1, along with some incrementing. This is called *parallel composition*; two Actions are composed in parallel to form a more complex Action. It is implemented with the natural hardware structure, shown in Figure 5.1:

- Q output of reg y1 goes through a +1 adder into D input of reg x1.

- Q output of reg x1 goes through a +1 adder into D input of reg y1.

The next example shows two rules which conflict—they can't be execute in the same clock:

```
rule r2a;
    $display ("r2a");
    x2 <= y2 + 1;
endrule

rule r2b;
    $display ("r2b");
    y2 <= x2 + 1;
endrule
```

In the logical semantics, either r2a executes before r2b or vice versa. Thus, from the initial state (x=10, y=100), we could have two different scenarios:

- if we executed r2a followed by r2b, the final state would be (x2=101, y2=102).

- if we executed r2b followed by r2a, the final state would be (x2=12, y2=11).

These are both legal final states of the rules in the logical semantics.

The hardware of Fig. 5.1 would not be a legal implementation of these logical semantics, since the final state after a clock would be (101, 11), which is not one of the two legal final states. In particular, the reads and writes in the logical semantics are reordered in the hardware in such a way as to produce different results.

Thus, we say that these two rules conflict. In fact, when we compile, we get an urgency warning message from the compiler:

```
Warning: "Tb.bsv", line 16, column 8: (G0010)
   Rule "r2b" was treated as more urgent than "r2a". Conflicts:
     "r2b" cannot fire before "r2a": calls to y2.write vs. y2.read
     "r2a" cannot fire before "r2b": calls to x2.write vs. x2.read
Warning: "Tb.bsv", line 30, column 9: (G0021)
   According to the generated schedule, rule "r2a" can never fire.
```

This warning is telling us that both rules cannot fire in the same clock because they conflict. Thus, in each clock, one rule has to be chosen; since the user has not specified this choice, the compiler arbitrarily picked rule r2b over r2a. Further, since both rules are enabled on every clock, and r2b is always preferred over r2a, the rule r2a can never fire. The next example demonstrates how to use scheduling attributes to specify which rule the compiler should choose in case of conflict.

5.3.1 Attribute `descending_urgency`

Scheduling attributes are used to control decisions made by the compiler. This attribute tells the compiler which rule is more urgent, that is which rule to schedule first when two rules conflict. Try adding this line before the rules:

```
(* descending_urgency = "r2a, r2b" *)
```

Now you'll see that the *more urgent* warning goes away, because now you have specified an urgency. The second warning (*can never fire*) will still be present, but it will now tell you that rule r2b can never fire, since the urgency specified in the attribute (r2a over r2b) is the opposite of what the compiler previously picked.

5.4   Multiple registers in a rigid synchronous pipeline

Source directory: `rulesregs/reg_pipeline`

Source listing: Appendix A.4.3

Now we'll look at a more complex example with multiple registers: a rigid synchronous pipeline. The example contains two modules: `mkTb`, the testbench module, and `mkPipe`, the pipeline module. Let's look at the testbench module first.

The testbench module instantiates a register with an initial value of 'h10, then instantiates the `mkPipe` module binding its interface to variable named `pipe` of type `Pipe_ifc`:

```
(* synthesize *)
module mkTb (Empty);

   Reg#(int) x     <- mkReg ('h10);
   Pipe_ifc  pipe <- mkPipe;
```

There are two rules in the testbench: `fill` and `drain`. The rule `fill`, which has no explicit rule condition, sends the value of x into the pipe and then increments it by 'h10.

```
   rule fill;
      pipe.send (x);
      x <= x + 'h10;
   endrule
```

The rule `drain`, which also has no explicit condition, receives the value y from the pipe and displays it. The program stops when y > 'h80.

```
   rule drain;
      let y = pipe.receive();
      $display ("    y = %0h", y);
      if (y > 'h80) $finish(0);
   endrule
endmodule
```

The first line in the above rule declares a local variable y and initializes it to the result of the statement `pipe.receive()`. The line could also have been written:

```
      int y = pipe.receive();
```

explicitly declaring y by giving its type, int. When we use let instead, the compiler infers the type for y, which must be int because that is the type of the right-hand side.

Now let's examine the module mkPipe, providing the Pipe_ifc interface with send (write) and receive (read) methods.

```
interface Pipe_ifc;
   method Action send (int a);
   method int    receive ();
endinterface
```

The pipeline module instantiates 4 int registers. mkRegU makes a register *uninitialized*, i.e., without specifying a reset value. The BSV compiler typically uses 'haaaaaaaa as the initial value for such registers (because of the 101010... pattern).

```
(* synthesize *)
module mkPipe (Pipe_ifc);

   Reg#(int) x1    <- mkRegU;
   Reg#(int) x2    <- mkRegU;
   Reg#(int) x3    <- mkRegU;
   Reg#(int) x4    <- mkRegU;
```

The rule r1 makes it a rigid synchronous pipeline, a rigid shift register. On each rule firing, each register is incremented and shifted into the next register. These shifts all happen simultaneously.

```
      rule r1;
         x2 <= x1 + 1;
         x3 <= x2 + 1;
         x4 <= x3 + 1;
      endrule
```

This is a parallel composition of three Actions into a single composite Action which forms the body of the rule.

Whenever the parent module (mkTb) invokes send, the int argument is stored in x1:

```
      method Action send (int a);
         x1 <= a;
      endmethod
```

Whenever the parent module invokes receive, the contents of x4 are returned:

```
    method int receive ();
       return x4;
    endmethod
```

The full code is found in A.4.3. When you execute the program, you'll see the pipeline shift register behavior. In the first few clocks, before the pipeline fills up, you'll see that the values being shifted and the value returned are the default reset values in the registers.

5.5 Using register syntax shorthands

Source directory: `rulesregs/reg_shorthand`

Source listing: Appendix A.4.4

In this section we'll modify the previous example to use register shorthands. Therefore, instead of the `mkPipe` module providing the `Pipe_ifc` interface, it will provide a `Reg#(int)` interface. When we instantiate the `mkPipe` module, we bind its interface to a variable named `pipe` of type `Reg#(int)` (instead of type `Pipe_ifc` in the previous example).

```
    Reg#(int) pipe <- mkPipe;
```

Registers have a special notation for the `_read` and `_write()` methods. Instead of explicitly using the methods, you can use the traditional non-blocking assignment notation (<=) for writes. For reads, you just mention the register interface in an expression.

Let's look at how we could replace the **send** and **receive** methods with register writes and reads in the `mkTb` module. To write the value of x to the pipe, we use the shorthand for `pipe._write(x)`:

```
    pipe <= x;     // was pipe.send(x)
```

To read a value out of `pipe`, we use the shorthand for `let y = pipe._read();`:

```
    let y = pipe;    // was let y = pipe.receive();
```

Instead of providing the `Pipe_ifc` interface, the `mkPipe` module provides a standard register interface:

```
    module mkPipe (Reg#(int));
```

The standard register interface provides the _write and _read methods instead of the send and receive methods in the previous example:

```
        method Action _write (int a);   // was method Action send (int a);
        method int _read ();            // was method int receive ();
```

5.6   Using valid bits

Source directory: rulesregs/validbits

Source listing: Appendix A.4.5

Let's expand the previous example, adding Valid bits to deal with pipeline bubbles. The testbench module, mkTb is not changed at all; all changes are in the mkPipe module.

In the mkPipe module, we instantiate 4 validn registers in addition to the xn registers:

```
      Reg#(Bool) valid1 <- mkReg(False);   Reg#(int) x1    <- mkRegU;
      Reg#(Bool) valid2 <- mkReg(False);   Reg#(int) x2    <- mkRegU;
      Reg#(Bool) valid3 <- mkReg(False);   Reg#(int) x3    <- mkRegU;
      Reg#(Bool) valid4 <- mkReg(False);   Reg#(int) x4    <- mkRegU;
```

where valid1 is a register holding a boolean value indicating whether x1 is valid. When you shift the value in the pipeline, shift the valid bit along with it:

```
        valid2 <= valid1;    x2 <= x1 + 1;
```

The function display_Valid_value displays Invalid or the valid value.

```
      function Action display_Valid_Value (Bool valid, int value);
         if (valid) $write (" %0h", value);
         else       $write (" Invalid");
      endfunction
```

The function type is Action; hence it can be invoked in the body of a rule. The _write and _read methods are written to take into account the valid bits:

```
      method Action _write (int a);
         valid1 <= True;
         x1 <= a;
      endmethod

      method int _read () if (valid4);
         return x4;
      endmethod
```

Because of the rule condition if (valid4), the _read method can only be called when valid4 is True.

Execute and observe the pipeline behavior. Further, note that until the 4-deep pipeline gets filled and produces a valid value from the register, no y value will be displayed by mkTb because the _read() method is not enabled until that time, therefore the drain rule cannot not fire until that time.

5.7 Multiple FIFOs in an elastic asynchronous pipeline

Source directory: rulesregs/fifo_pipeline

Source listing: Appendix A.4.6

This example takes the synchronous pipeline example presented in Section 5.4 and replaces the registers with FIFOs. To use the Bluespec-provided FIFO interface and module definitions, you must first import the FIFO package:

```
import FIFO::*;
```

Once again, there is no change to the mkTb module. In this example the mkPipe module provides a modified Pipe_ifc interface, in which the receive method is now an ActionValue method (instead of a Value method). An ActionValue method combines Value and Action methods, returning a value (in this case an int value) and performing an Action. Because of the latter, it can only be called in Action contexts, namely in the bodies of rules (or from within other Action or ActionValue methods).

```
interface Pipe_ifc;
   method Action            send (int x);
   method ActionValue#(int) receive;
endinterface
```

Instead of instantiating 4 registers, we instantiate 4 FIFOs (all initially empty) and bind the interfaces to f1...f4:

```
FIFO#(int) f1    <- mkFIFO;
FIFO#(int) f2    <- mkFIFO;
FIFO#(int) f3    <- mkFIFO;
FIFO#(int) f4    <- mkFIFO;
```

The rule r1 from the earlier example is replaced by the rules r2, r3, and r4 to read the FIFO values and move them through the pipeline.

```
rule r2;
    let v1 = f1.first;
    f1.deq;
    $display (" v1 = %0h", v1);
    f2.enq (v1+1);
endrule

rule r3;
    let v2 = f2.first;
    f2.deq;
    $display (" v2 = %0h", v2);
    f3.enq (v2+1);
endrule

rule r4;
    let v3 = f3.first;
    f3.deq;
    $display (" v3 = %0h", v3);
    f4.enq (v3+1);
endrule
```

The rules r2, r3, and r4 may or may not fire together. Each rule fires whenever its input FIFO has data available (implicit condition of `first` and `deq`) *and* its output FIFO has space available (implicit condition of `enq`).

The Action method `send` enqueues a value `a` into FIFO f1. This method is only enabled if f1 has space available. Thus the implicit condition of `f1.enq` becomes the implicit condition of this `send` method and becomes part of the enabling condition of the rule `fill` in the testbench, i.e., that rule cannot fire unless the implicit conditions of `f1.enq` are met.

```
method Action send (int a);
    f1.enq (a);
endmethod
```

The ActionValue method `receive` reads the first `int` element v4 of FIFO f4, dequeues FIFO f4 and returns v4. This method is only enabled when f4 has data available (implicit condition of `first` and `deq`). Thus the implicit conditions of `f4.first` and `f4.deq` become the implicit condition of the `receive` method, which in turn becomes part of the enabling condition of rule `drain` in the testbench. That is, the rule `drain` cannot fire until all these conditions are true.

```
method ActionValue#(int) receive ();
   let v4 = f4.first;
   $display (" v4 = %0h", v4);
   f4.deq;
   return v4;
endmethod
```

When executed, you can observe the pipeline behavior. However, the outputs from the $displays in different rules that execute in the same cycle may be unpredictable.

Chapter 6

Module hierarchy and interfaces

6.1 Module hierarchies

Source directory: `modules/mod_hierarchy`

Source listing: Appendix A.5.1

Modules and interfaces are the way we organize large, complex systems into smaller, manageable components. Modules and interfaces turn into actual hardware. Modules can *instantiate* other modules, resulting in a *module hierarchy*. Logically, the hierarchy forms a tree structure. A module consists mainly of three things: zero or more sub-modules, rules that operate on the sub-modules, and the module's interface to the outside world (the surrounding hierarchy).

Every module *provides* an interface, though this may be the `Empty` interface. An interface for a module m mediates between m and other, external modules that *use* the interface to interact with the facilities of m. We often refer to these other modules as *clients of m*.

A module's BSV interface is an abstraction of its Verilog port list. An interface provides a means to group wires into bundles with specified uses, described by methods.

By separating out the interface declaration from module declarations, we express the idea of a common interface that may be provided by several modules, without having to repeat that declaration in each of the implementation modules.

An interface declaration specifies the methods provided by every module that provides the interface, but does not specify the methods' implementation. The implementation of the interface methods can be different in each module that provides that interface. The definition of the interface and its methods is contained in the providing module.

It is important to distinguish between a module *definition* and a module *instantiation*. A module definition can be regarded as specifying a scheme that can be instantiated multiple times. For example, we may have a single module definition for a FIFO, and a particular design may instantiate it multiple times for all the FIFOs it contains. For this reason, we follow a convention where our modules have names like "mkFoo", with the "mk" prefix suggesting the work "make", because the module definition can be used to make multiple module instances.

When instantiated, the current example produces the module hierarchy shown in Figure 6.1. The figure on the right shows the interface provided by each instantiated module.

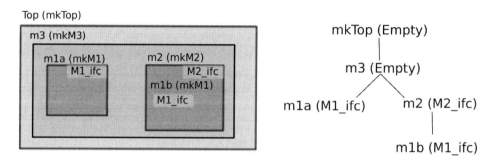

Figure 6.1: Module Hierarchy

Every module uses the interface(s) just below it in the hierarchy. Every module provides an interface to the module above it in the hierarchy. For example, in the hierarchy in Figure 6.1, the following interfaces are used and provided:

- Module m1a provides interface M1_ifc

- Module m2 provides interface M2_ifc

- Module m3 uses interface M2_ifc

- Module m1b provides interface M1_ifc

The top-level module mkTb, instantiates an instance m3 of the module mkM3:

```
package Tb;

(* synthesize *)
module mkTb (Empty);
   Empty m3 <- mkM3;
endmodule
```

The module mkM3 instantiates an instance of mkM1 and mkM2:

```
(* synthesize *)
module mkM3 (Empty);

   Reg#(int) x    <- mkReg (10);
   M1_ifc     m1b <- mkM1 (20);
   M2_ifc     m2  <- mkM2 (30);
   ...
endmodule
```

The module mkM2, instantiates another instance of mkM1:

```
module mkM2 #(parameter int init_val) (M2_ifc);

   M1_ifc  m1a <- mkM1 (init_val + 10);
```

6.1.1   Parameter

The statement:

```
module mkM3(Empty);
   ...
   M1_ifc     m1b <- mkM1 (20);
```

instantiates module mkM1 inside the module M3 with the static elaboration parameter 20; the interface provided by the instantiation is bound to the variable named m1b of type M1_ifc.

A module definition can include a parameter for the module. If the parameter keyword is used, then the parameter becomes a Verilog parameter and not a Verilog port (input wire). In our example, the definition of module mkM2 includes the parameter init_val:

```
module mkM2 #(parameter int init_val) (M2_ifc);
```

Wherever mkM2 is instantiated, the actual parameter must be a static (compile-time) value.

The statement:

```
   M1_ifc  m1a <- mkM1 (init_val + 10);
```

instantiates the module mkM1 inside the module M2, passing it a parameter that is mkM2's parameter + 10.

6.1.2   Scope of methods

The statement:

```
$display ("x = %0d, m1b/x = %0d, m2/x = %0d, m2/m1a/x = %0d",
          x, m1b.read_local_x, m2.read_local_x, m2.read_sub_x);
```

displays the value of x in each module instance, using method invocations to access x in the sub-modules.

Note that there are 4 registers called x, one at each level of the module hierarchy (x, m1b.x, m2.x, m2.m1.x) even though there are only three textural instances of mkReg:

```
Reg#(int) x <- mkReg (init_val);
```

Because the module mkM1 is instantiated twice, we get 4 registers after static elaboration.

Modules mkM1 and mkM2 each have their own definition of the method read_local_x. These methods both return a value of x as well.

```
interface M2_ifc;
   method int read_local_x ();
   method int read_sub_x ();
endinterface

module mkM2 #(parameter int init_val) (M2_ifc);
   M1_ifc  m1a <- mkM1 (init_val + 10);
   Reg#(int) x <- mkReg (init_val);

   method int read_local_x ();
      return x;
   endmethod
```

```
interface M1_ifc;
   method int read_local_x ();
endinterface

(* synthesize *)
module mkM1 #(parameter int init_val) (M1_ifc);

   Reg#(int) x <- mkReg (init_val);

   method int read_local_x ();
       return x;
   endmethod

endmodule: mkM1
```

There are 2 definitions of methods called `read_local_x`, but three invocations:

- m3 invoking m1b.read_local_x

- m3 invoking m2.read_local_x

    - m2 invoking m1a.read_local_x

BSV uses standard programming language *scope* rules to resolve identifiers, so there is no ambiguity about the different uses of x or `read_local_x`.

6.1.3   Analyzing a design

The design browsers within the Bluespec Development Workstation provide many views to facilitate your understanding and analysis of a design. The following table summarizes the windows and browsers in the workstation. Which browser you use for a given task depends on the information you want to see and the state of the design.

Bluespec Development Workstation Windows		
Window	Function	Required Files
Main Window	Central control window. Manage projects, set project options, build projects, and monitor status.	.bspec
Project Files Window	View, edit and compile files in the project.	.bsv
Package Window	Pre-elaboration viewer for the design. Load packages into the workstation and browse their contents. Provides a high-level view of the types, interfaces, functions and modules defined in the package.	.bi/.bo
Type Browser	Primary means for viewing information about types and interfaces. Displays the full structure hierarchy and all the concrete types derived from resolution of polymorphic types.	.bi/.bo
Module Browser	Post-elaboration module viewer, including rules and method calls. Displays the design hierarchy and an overview of the contents of each module. Provides an interface to the waveform viewer.	.ba
Schedule Analysis Window	View schedule information including warnings, method calls, and conflicts between rules for a module.	.ba
Scheduling Graphs	Graphical view of schedules, conflicts, and dependencies.	.ba .dot

To view the module hierarchy for the example in this section, open the file `modules/mod_hierarchy/mod_hierarchy.bspec`. Compile the design, open the **Module Browser**, and **Load Top Module** to display the module hierarchy. Once you've simulated the design, from this window you can link to an external waveform viewer, using the instantiated module hierarchy and type definitions along with waveform dump files to display Bluespec type data along with the waveforms.

To view waveforms, you must generate a waveform dump file, either VCD or FSDB, during simulation. You can generate a VCD file from either Bluesim or any of the supported Verilog simulators. When simulating with Bluesim, use the `-V` flag. For Verilog simulators using the Bluespec-provided `main.v` file, specify the `+bscvcd` flag during simulation. Simulation flags are entered in the options field of the **Project Options→Link/Simulate** window.

More information on the development workstation and Bluesim modes is described in the *Bluespec SystemVerilog User Guide*.

6.2 Implicit conditions of methods

Source directory: `modules/implicit_cond`

Source listing: Appendix A.5.2

This example contains two methods which are very similar:

```
interface Dut_ifc;
   method int f1 ();
   method int f2 ();
endinterface

(* synthesize *)
module mkDut (Dut_ifc);
   ...
   method int f1 ();
      return x;
   endmethod

   method int f2 () if (is_even(x));
      return x;
   endmethod

endmodule
```

The only difference between the two methods is that `f2` has an *implicit condition*; `f2` can only be invoked when `x` is even. The implicit condition may be a boolean expression or it may be a pattern-match, but cannot include updating of state elements. A method can only fire when the implicit condition is `True` or matches the pattern. Expressions in the implicit condition can use any of the variables in scope surrounding the method definition, i.e., visible in the module body, but they cannot use the formal parameters of the module itself. If the implicit condition is a pattern-match, any variables bound in the pattern are available in the method body. Omitting the implicit condition is equivalent to saying `if (True)`.

The testbench executes two rules which look very similar, almost identical, the difference in execution being the implicit condition on `f2`.

```
      rule r1;
         let x1 = dut.f1;
         ...
      rule r2;
         let x2 = dut.f2;
         ...
```

When executed, one can observe that rule `r1` fires on every clock and rule `r2` fires on every other clock because of the implicit condition on method `f2`.

6.3   ActionValue methods

Source directory: `modules/actionvalue`

Source listing: Appendix A.5.3

An ActionValue statement performs an Action (update of state element) and returns a value. In this section we're going to look at the statement:

```
int w <- dut.avmeth(10);
```

This statement invokes the `avmeth` ActionValue method with an argument 10, causing an action inside the module. The method returns a value which is assigned to the local variable `w` of type `int`.

6.3.1   The let statement

This statement could also have been written with a `let`:

```
let w <- dut.avmeth(10);
```

The `let` statement is a shorthand way to declare and initialize a variable in a single statement. A variable which has not been declared can be assigned an initial value and the compiler will infer the type of the variable from the expression on the right hand side of the expression.

If the expression on the right hand side is the value returned by an ActionValue method, use the `<-` assignment operator. For Value methods, use the `=` assignment operator.

6.3.2   The <- operator

Technically, the type of the expression `dut.avmeth(10)` is `ActionValue#(int)`, not `int`. If we had used `=` instead of `<-`, then the compiler would produce a type error, since it is an attempt to assign a value of type `ActionValue#(int)` to a variable of type `int`. Section 6.4 demonstrates what happens when you use the wrong operator.

When using `<-`, the right-hand side must be an expression of type `ActionValue#(t)` for some type `t`. The statement causes the Action to be performed, and assigns the returned value to the variable on the left-hand side, which must be of type `t`.

6.3.3   Defining the ActionValue Method

The statement:

```
interface Dut_ifc;
    method ActionValue#(int) avmeth (int v);
endinterface
```

declares `avmeth` to be an ActionValue method with an argument of type `int` that returns a value of type `int`. The declaration is found in the interface definition. The body of the method, describing how to accomplish the method, is defined in the module:

```
module mkDut (Dut_ifc);

   Reg#(int) x <- mkReg (0);

   method ActionValue#(int) avmeth (int v);
      x <= x + v;
      return x;
   endmethod

endmodule
```

The definition indicates that the method performs an action (incrementing register x by the value of v) and returns a value (the old value of x). When executed we can see that on each clock the current value of the register is printed, and it is updated by 10.

6.4 ActionValue method with type error

Source directory: `modules/action_value_type_error`

Source listing: Appendix A.5.4

This example is identical to Section 6.3, except for the deliberate error:

```
module mkTb (Empty);
   Dut_ifc dut <- mkDut;

   rule r1;
      int w = dut.avmeth(10);        // DELIBERATE ERROR
      $display ("w = %0d", w);
      if (w > 50) $finish (0);
   endrule
endmodule
```

The right-hand side expression, `dut.avmeth(10)` has a type of `ActionValue#(int)`. This statement attempts to bind this to a variable of type `int`, using an `=` operator instead of the `<-` operator, provoking the following type error:

```
Error: "Tb.bsv", line 20, column 15: (T0080)
   Type error at the use of the following function:
```

```
dut.avmeth
```

The expected return type of the function:
```
Int#(32)
```

The return type according to the use:
```
ActionValue#(Int#(32))
```

6.5 ActionValue method with error in rule condition

Source directory: `modules/action_value_rule_error`

Source listing: Appendix A.5.5

Rules and method definitions can have conditions. A rule condition may be either a boolean expression or a pattern-match. The condition can use any identifiers from the scope surrounding the rule, i.e. visible in the module body. If the rule condition is a pattern-match, any variables bound in the pattern are available in the rule body. Most importantly, expressions in the rule conditions cannot have side effects. In general, any Action or ActionValue expression represents a side effect, that is, a potential change of state. Rule conditions and method implicit conditions can never contain an Action or an ActionValue method.

This example, has the deliberate error of defining a rule condition that contains an ActionValue:

```
module mkTb (Empty);
   Dut_ifc dut <- mkDut;

   rule r1 (dut.avmeth (10) > 0);     // DELIBERATE ERROR
      int w = dut.avmeth(10);
      $display ("w = %0d", w);
      if (w > 50) $finish (0);
   endrule

endmodule
```

An attempt to compile this will provoke the following type-checking error:

```
Error: "Tb.bsv", line 19, column 29: (T0031)
  The provisos for this expression could not be resolved because there are no
  instances of the form:
    Ord#(ActionValue#(Int#(32)))
```

This error may seem a bit cryptic at first, but let's look at it in more detail. The expression `dut.avmeth (10) > 0` is trying to use the greater than operator (`>`) to compare two values. In general, `>` is defined only on types in the `Ord` typeclass. Here it is complaining that `ActionValue#(int)` is not in the Ord typeclass, and hence can't apply the > operator to such an expression. Note: int is synonymous with Int#(32), so ActionValue#(int) is the same as ActionValue#(Int#(32)).

As described in Section 3.3, overloading is the ability to use a common function name or operator on a collection of types. For example, the < function is meaningful on types that can be compared such as bits, numeric types, or vectors. The implementation of how you compare the various types is different, but all the members of the Ord typeclass allow the comparison. In this example, we have an ActionValue, which is not a type that can be compared with anything. An ActionValue type is not a member of the Ord typeclass.

6.6 Defining an Action function

Source directory: modules/action_function

Source listing: Appendix A.5.6

Previously we've discussed Action methods, which are defined as part of an interface. In this section, we're going to examine Action *functions*, which are not part of an interface definition. Any expression that is intended to act on the state of the circuit (at circuit execution time) is an *action* and has type Action. Actions are combined by using action blocks, with a action-endaction structure.

We're going to write an Action function that represents the incrementing of x by a value dx, and the incrementing of y by a value dy. Rules represent atomic transactions. Rule bodies specify the collection of Actions that are performed atomically when the rule executes. This function constructs and returns an Action, using the action-endaction block that can be used in a rule body, in the same way that an Action method constructs an Action that can be used in a rule body. However, unlike an Action method, an Action function is not part of an interface, does not explicitly specify any condition, and is just considered an abbreviation of the Actions in the function body. In short, functions allow you to construct parameterized Actions.

```
module mkTb (Empty);

    Reg#(int) x <- mkReg (0);
    Reg#(int) y <- mkReg (0);

    function Action incr_both (int dx, int dy);
      return
        action
            x <= x + dx;
            y <= y + dy;
        endaction;
    endfunction
```

This function is defined inside the module as a local function. It is not visible at all outside this module. Functions can also be defined at the top level, inside other functions, and, in general, in any scope. Normal scoping rules apply. For example, the use of x and y inside the function body refer to the registers x and y in the function's surrounding scope, namely in the module Tb.

The `incr_both` function is invoked in both rule r1 and rule r2.

```
rule r1 (x <= y);
   incr_both (5, 1);
   $display ("(x, y) = (%0d, %0d); r1 fires", x, y);
   if (x > 30) $finish (0);
endrule

rule r2 (x > y);
   incr_both (1, 4);
   $display ("(x, y) = (%0d, %0d): r2 fires", x, y);
endrule
```

The semantics are directly equivalent to inlining the function where it is invoked, and binding the function arguments correctly: (5,1) to (dx,dy) in the rule r1 and (1,4) to (dx,dy) in the rule r2.

6.7   Nested interfaces

Source directory: `modules/nested_ifc`

Source listing: Appendix A.5.7

Interfaces can be defined hierarchically; that is an interface can be defined in terms of other interfaces or subinterfaces. This can be a valuable design technique, especially when you define reusable subinterfaces.

A common paradigm between two blocks is a get/put; one side *gets* or retrieves an item from an interface, the other side *puts* or gives an item to an interface. These gets and puts can be combined to define clients and servers.

Let's define two interfaces: a `Client_int` interface, which *gets* requests and *puts* responses, and a `Server_int` interface which *puts* requests and *gets* responses. Instead of defining methods in these interfaces directly, we are going to define them hierarchically. First we'll declare the interfaces `Get_int` and `Put_int`, because these interfaces capture a common, reusable paradigm. We'll then use the `Get_int` and `Put_int` interfaces as subinterfaces of the `Client_int` and `Server_int` interfaces.

The `Put_int` interface has a `put` Action method which gives an item of type `int` to an interface.

```
interface Put_int;
   method Action put (int x);
endinterface
```

The `Get_int` interface has a `get` ActionValue method, similar to a dequeue, which retrieves an item for the interface and removes it at the same time.

```
interface Get_int;
    method ActionValue#(int) get ();
endinterface
```

We can combine these interfaces to declare the client and server interfaces. The `Client_int` has a subinterface to get requests and another subinterface to put responses. Conversely, the `Server_int` has a subinterface to put requests and another subinterface to get responses.

```
interface Client_int;
    interface Get_int  get_request;
    interface Put_int  put_response;
endinterface: Client_int

interface Server_int;
    interface Put_int  put_request;
    interface Get_int  get_response;
endinterface: Server_int
```

Now let's look at how we can use these interfaces in an working design. In this example we have a stimulus generator (`mkStimulusGen`) providing a `Client_int` interface, a DUT (`mkDut`) providing a `Server_int` interface, and a top level module (`mkTb`) connecting them together.

The top-level module instantiates the two submodules (the stimulus generator and the DUT) and connects the flow of requests from the stimulus generator to the DUT and the flow of responses from the DUT to the generator using two rules.

```
module mkTb (Empty);

    Client_int  stimulus_gen <- mkStimulusGen;
    Server_int  dut          <- mkDut;

    // Connect flow of requests from stimulus generator to DUT
    rule connect_reqs;
        let req <- stimulus_gen.get_request.get();
        dut.put_request.put (req);
    endrule

    // Connect flow of responses from DUT back to stimulus generator
    rule connect_resps;
        let resp <- dut.get_response.get();
        stimulus_gen.put_response.put (resp);
    endrule
endmodule: mkTb
```

Let's examine the rule connect_reqs to see how the nested interfaces are used to connect the request from the client to the request of the server. We know that stimulus_gen has the type Client_int, which has two subinterfaces: get_request and put_response. The first statement assigns the value returned by the method get from the get_request interface to the variable req. The second statement, selects the subinterface put_request from the Server_int interface (since dut has type Server_int), and selects the put method. It will put the value received from the get.

The rule connect_resps similarly connects the response from the DUT back to the stimulus generator. The first statement *gets* the response and *puts* and in the second statement that value is *put* to the stimulus generator.

The definitions of the interface methods are defined within the modules; the Client_int methods are defined in the module mkStimulusGen and the Server_int methods are defined in the module mkDut. Interfaces define a common set of wires and methods; how these methods are implemented are defined within the module providing the interface. Different modules can have different method definitions for a common interface.

First let's look at the mkDut module which provides a Server_int interface.

```
module mkDut (Server_int);

   FIFO#(int) f_in  <- mkFIFO;     // to buffer incoming requests
   FIFO#(int) f_out <- mkFIFO;     // to buffer outgoing responses

   rule compute;
      let x = f_in.first; f_in.deq;
      let y = x+1;                        // Some 'server' computation (here: +1)
      if (x == 20) y = y + 1;            // Modeling an 'occasional bug'
      f_out.enq (y);
   endrule

   interface Put_int put_request;
      method Action put (int x);
          f_in.enq (x);
      endmethod
   endinterface

   interface Get_int get_response;
      method ActionValue#(int) get ();
          f_out.deq; return f_out.first;
      endmethod
   endinterface

endmodule: mkDut
```

Let's look how the `Server_int` get and put methods are defined. There are two ways of defining the methods in the interface: directly and by assignment. Here we define the method in the `put_request` subinterface using the *direct* syntax. Whenever the `put` method inside this subinterface is called, it enqueues the argument onto the FIFO `f_in`.

The `get_response` interface subinterface also uses the direct definition syntax. When the `get` method is called, it dequeues the value from `f_out` and returns the `first` value from `f_out`.

Now let's examine the `mkStimulusGen` module.

```
module mkStimulusGen (Client_int);

   Reg#(int)  x            <- mkReg (0);        // Seed for stimulus
   FIFO#(int) f_out        <- mkFIFO;           // To buffer outgoing requests
   FIFO#(int) f_in         <- mkFIFO;           // To buffer incoming responses
   FIFO#(int) f_expected   <- mkFIFO;           // To buffer expected responses

   rule gen_stimulus;
      f_out.enq (x);
      x <= x + 10;
      f_expected.enq (x+1);     // mimic functionality of the Dut 'server'
   endrule

   rule check_results;
      let y     = f_in.first; f_in.deq;
      ...
   endrule
```

In this module we use the define the `put_response` subinterface directly and the `get_request` subinterface by assignment.

```
   interface get_request = interface Get_int;
                              method ActionValue#(int) get ();
                                 f_out.deq; return f_out.first;
                              endmethod
                           endinterface;

   interface Put_int put_response;
      method Action put (int y) = f_in.enq (y);
   endinterface
```

The entire `interface...endinterface` phrase after the `=` is an *interface expression*, that is, an expression whose value is an interface. This value is assigned to the `get_request` interface variable on the left-hand side. Why is this useful? First, it allows for flexible parameterized static elaboration. For example, we could include an `if...else` statement:

```
   interface _get_request = if (...static_parameter...)
                              interface Get_int ... ... endinterface
                           else
                              interface Get_int ... <def> ... endinterface
```

or even

```
interface _get_request = ... function returning an interface ...
```

The following exercises will help you understand interfaces and subinterfaces in more detail.

1. Examine the Verilog generated for these modules and understand the correspondence between the BSV interface and method definitions and the Verilog module port lists.

2. In mkTb, we have two rules: connect_reqs and connect_resps, one for each direction. What happens if we combine them into a single rule?

3. Execute the program and note the cycles in which the rules fire. For the FIFO f_expected, replace the mkFIFO constructor with mkSizedFIFO(10). Observe again the cycles in which rules fire. Why are they different? Hint: the mkFIFO constructor creates a default-sized FIFO (usually 2 elements), whereas the mkSizedFIFO(10) constructor creates a FIFO with 10 elements. Instead of 10, how small can the FIFO be without changing behavior?

4. In the rule check_results, what would happen if we forgot to write the statement f_expected.deq?

6.8 Standard connectivity interfaces

Source directory: modules/connect_ifc

Source listing: Appendix A.5.8

The example in the previous section had interfaces for *getting an item* and *putting an item* and a top-level module connecting them. These types of interfaces are often called *Transaction Level Modeling* or *TLM* for short. This pattern is so common in system design that BSV provides parameterized library interfaces, modules, and functions for this purpose. This example implements the previous example using library elements instead of defining them from scratch. It reflects a common paradigm in BSV coding, using BSV's powerful library elements to quickly put together models and designs.

When using elements from the BSV library, you need to *import* the packages containing the components you need. We're going to import the BSV libraries for FIFOs, Get and Put interfaces, Client and Server interfaces, and mkConnection modules.

```
import FIFO::*;
import GetPut::*;
import ClientServer::*;
import Connectable::*;
```

The mkTb module in this example will use the BSV-defined Client and Server interfaces, replacing the Client_int and Server_int interfaces in the previous example.

```
    module mkTb (Empty);
        Client#(int,int) stimulus_gen <- mkStimulusGen;
        Server#(int,int) dut           <- mkDut;
```

The type `Client#(req_t, resp_t)` is a client interface yielding requests of type `req_t` via a `Get` subinterface and accepting responses of type `resp_t` via a `Put` subinterface.

```
interface Client#(type req_type, type resp_type);
    interface Get#(req_type)  request;
    interface Put#(resp_type) response;
endinterface: Client
```

The type `Server#(req_t, resp_t)` is a server interface accepting requests of type `req_t` via a `Put` subinterface and yielding responses of type `resp_t` via a `Get` interface.

```
interface Server#(type req_type, type resp_type);
    interface Put#(req_type)  request;
    interface Get#(resp_type) response;
endinterface: Server
```

The package `Connectable` defines an overloaded module called `mkConnection`, which is used to connect a client to a server.

```
    mkConnection (stimulus_gen, dut);
```

This connects a Client to a Server in both directions. Technically, it instantiates a module that takes these two interfaces as parameters. The module implements the functionality of the rules from the previous example:

- a rule to get a request from the client and `put()` it into the server (`connect_reqs`)

- a rule to get a response from the server and `put()` it into the client (`connect_resps`).

You may ask, if `mkConnection` is instantiating a module, then what is its interface? `mkConnection` has been defined with an `Empty` interface, so technically, the statement could have been written in full as follows:

```
    Empty e <- mkConnection (stimulus_gen, dut);
```

Since there is no subsequent use for the `Empty` interface e, we are allowed to write this in the shorter form shown in the actual code.

The `mkTb` module is simplified from the previous example, using `mkConnection` instead of the rules `connect_reqs` and `connect_resps`:

```
module mkTb (Empty);

    Client#(int,int) stimulus_gen <- mkStimulusGen;
    Server#(int,int) dut          <- mkDut;

    mkConnection (stimulus_gen, dut);
endmodule: mkTb
```

The state and rules of `mkDut` are the same as in the previous example, but here we are defining the subinterfaces `request` and `response` using the assignment syntax.

```
module mkDut (Server#(int,int));
    ...
    interface request  = toPut (f_in);
    interface response = toGet (f_out);
```

The function `toPut()` is a library function which we sometimes call an *interface transformer*; it takes a FIFO interface argument (here, `f_in`) and returns a `Put` interface result. Similarly, `toGet()` is a library interface transformer that transforms from a FIFO to a `Get`. There is nothing magical about these interface transformers; they are ordinary BSV functions that you could write yourself.

In fact, here is another example of such an interface transformer, `fifosToClient`. This function, written for this example, combines the two interface transformers.

```
function Client#(req_t, resp_t) fifosToClient (FIFO#(req_t) f_reqs,
                                               FIFO#(resp_t) f_resps);
    return interface Client
             interface Get request  = toGet (f_reqs);
             interface Put response = toPut (f_resps);
           endinterface;
endfunction: fifosToClient
```

The `fifostoClient` transformer makes the interface definition even more succinct, a one-liner, used in the `mkStimulusGen` module.

```
module mkStimulusGen (Client#(int,int));
    ...
    return fifosToClient (f_out, f_in);
```

We could have defined the interface in `mkStimulusGen` using two lines, like this:

```
          interface request   = toGet (f_out);
          interface response = toPut (f_in);
```

Instead, we directly build the entire interface, which includes the two subinterfaces, by calling the interface transformer, `fifosToClient()`. This is just a function that takes two `FIFO` interfaces and returns a `Client` interface that gets requests from the first FIFO and puts responses into the second FIFO. In addition, note that it is polymorphic (described in more detail in Section 9). That is, it works for any request type `req_t` and any response type `resp_t`.

Using polymorphism, parameterization, and higher-order programming, BSV allows you to define very powerful and highly reusable components, including types, interfaces, modules, and functions.

Chapter 7

Scheduling

As described in Section 5.2, state updates are defined within rules. While each rule execution is logically atomic, and ordered with respect to all other rule executions, the actual behavior in hardware is that multiple rules are scheduled concurrently. The BSV design is mapped by the compiler into parallel clocked synchronous hardware. The mapping permits multiple rules to be executed in each clock cycle.

The BSV compiler includes a dynamic scheduler that allows many rules to fire in parallel in each clock. The compiler performs a detailed and sophisticated analysis of the rules and their interactions and maps the design into very efficient, highly parallel, clocked synchronous hardware. The designer can use the simple reference (one-at-a time) semantics to reason about correctness properties and be confident that the synthesized parallel hardware will preserve those properties. BSV also provides scheduling attributes to allow the designer to guide the compiler in implementing the mapping.

While respecting the constraints of atomicity and rule conflicts, there are still many choices in which subsets of rules can fire in each clock cycle. A central principle is that BSV's scheduling is "greedy":

- Every rule enabled during a clock cycle will fire unless it is prevented by a conflict. Rules that may not fire due to conflicts are reported as warnings during compilation (unless overridden by an explicit user-supplied scheduling attribute).

Conflicts are discussed in more detail in Section 5.2. Two or more rules conflict either because they are competing for a limited resource or because the result of their simultaneous hardware execution would be inconsistent with any sequential (atomic) rule execution. In the absence of a user annotation (attribute), the compiler will arbitrarily choose which rule to prioritize, but will also issue an informational warning[1]. This guarantees the designer is aware of the ambiguity in the design and can correct it. It might be corrected by changing the rules (rearranging their predicates so they are never simultaneously applicable, for example) or by adding an urgency attribute which

---

[1]The compiler's choice, while arbitrary, is deterministic. Given the same source and compiler version, the same schedule (and, hence, the same hardware) will be produced. However, because it is an arbitrary choice, it can be sensitive to otherwise irrelevant details of the program and is not guaranteed to remain the same if the source or compiler version changes.

tells the compiler which rule to prefer. When there are no scheduling warnings, it is guaranteed that the compiler is making no arbitrary choices about which rules to execute.

Another scheduling principle is that a rule will fire at most once during a particular clock cycle. This ensures that continuously enabled rules (like a counter increment rule) will only be executed at most one time during a clock cycle.

Together, these two principles allow a designer to understand precisely which rules fire during each cycle, and why; this, in turn, allows him to reason about performance.

7.1   Scheduling error due to parallel composition

Source directory: `schedule/parallel_error`

Source listing: Appendix A.6.1

This example consists of two modules: `mkTb1` and `mkTb`. In both modules, a FIFO is instantiated, `mkTb1` tries to invoke the `enq` method twice from the same rule (resulting in an error), while `mkTb` invokes the `enqs` from separate rules.

In `mkTb1`, the `enq` rule tries to enq from the same FIFO twice, from within the same rule:

```
mkTb1 (Empty);

   FIFO#(int) f <- mkFIFO;     // Instantiate a FIFO
   ...
   rule enq (state < 7);
      f.enq(state);
      f.enq(state+1);
      $display("FIFO enq: %d, %d", state, state+1);
   endrule
```

Since the `enq` method of a FIFO can only be invoked once per clock, the two `enq` methods conflict; they are not "parallel composable". The compilation fails with the error:

```
Error: "Tb.bsv", line 36, column 9: (G0004)
  Rule 'RL_enq' uses methods that conflict in parallel:
    f.enq(...)
  and
    f.enq(...)
  For the complete expressions use the flag '-show-range-conflict'.
```

Alternatively, if you comment out the `mkTb1` module, you will note that `mkTb` is a similar module, but it invokes the two `enq` methods from two separate rules:

```
module mkTb (Empty);

   FIFO#(int) f <- mkFIFO;     // Instantiate a FIFO
   ...
   rule enq1 (state < 7);
      f.enq(state);
      $display("FIFO enq: %d", state);
   endrule

   rule enq2 (state > 4);
      f.enq(state+1);
      $display("FIFO enq: %d", state+1);
   endrule
```

These two rules conflict for some cycles, since they both invoke f.enq the compiler cannot schedule them to fire in parallel. However, this will **not** cause a compilation error, instead the compiler will issue the warning:

```
Warning: "Tb.bsv", line 48, column 8: (G0010)
   Rule "enq2" was treated as more urgent than "enq1". Conflicts:
     "enq2" cannot fire before "enq1": calls to f.enq vs. f.enq
     "enq1" cannot fire before "enq2": calls to f.enq vs. f.enq
```

Note that when executing, enq1 can only fire when enq2 is not enabled by the rule condition (state <=4). When the rule conditions have both rules enabled (state 5 and 6), only enq2 fires.

7.2   Scheduling attributes

Scheduling attributes are used to express performance requirements. When the compiler maps rules into clocks, scheduling attributes guide or constrain its choices in order to produce a schedule that will meet performance goals.

Scheduling attributes are most often attached to rules or rule expressions, but some can also be added to module definitions. The attributes are only applied when the module is synthesized.

7.2.1   Prioritization using descending_urgency

Source directory: schedule/desc_urgency

Source listing: Appendix A.6.2

When the compiler maps rules into clocks the order in which rule conditions are considered can impact the subset chosen. For example, suppose rules r1 and r2 conflict, and both their conditions are true so both can execute. If r1 is considered first and selected,it will disqualify r2 from consideration, and vice versa.

The `descending_urgency` attribute allows the designer to specify that one rule is more *urgent* than another, so that it is always considered for scheduling before the other. The relationship is transitive, i.e. if rule `r1` is more urgent that rule `r2` and rule `r2` is more urgent than rule `r3`, then `r1` is more urgent than `r3`. If the urgency attributes are contradictory (resulting in an urgency cycle), the compiler will report an error.

The argument of the `descending_urgency` attribute is a string containing a comma-separated list of rule names. Example:

```
(* descending_urgency = "r1, r2, r3" *)
```

The `descending_urgency` attribute can be placed in one of three syntactic positions:

- It can be placed just before the `module` keyword in a module definition, in which case it can refer directly to any of the rules inside the module.

- It can be placed just before the `rule` keyword in a rule definition, in which case it can refer directly to the rule or any other rules at the same level.

- It can be placed just before the `rules` keyword in a rules expression, in which case it can refer directly to any of the rules in the expression.

In the previous example, the compiler issued a warning when `mkTb` is compiled, stating that `enq2` was treated as more urgent than `enq1`.

Adding the `descending_urgency` attribute will guide the compiler as to which rule it should treat as more urgent and give higher priority when both rules can fire.

```
(* descending_urgency = "enq1, enq2"*)
rule enq1 (state < 7);
   f.enq(state);
   $display("FIFO enq1: %d", state);
endrule

rule enq2 (state > 4);
   f.enq(state+1);
   $display("FIFO enq2: %d", state+1);
endrule
```

With this example, the compiler will schedule rule `enq1` to fire when both rules can fire (when state = 5 or 6).

7.2.2 Descending urgency vs. execution order

Source directory: `schedule/exec_order`

Source listing: Appendix A.6.3

We often use the terms execution order, earliness, and TRS order (from Term Rewriting Systems) synonymously. All these refer to the logical order in which rules execute within a clock cycle such that all methods are called in an order consistent with their scheduling constraints.

The details of the testbench in this example are not important. The focus is on the module `mkGadget` and the schedule of its rules. The module has two FIFOs: `infifo` and `outfifo`:

```
module mkGadget (Server#(int,int));
    int bubble_value = 42;

    FIFO#(int) infifo <- mkFIFO;
    FIFO#(int) outfifo <- mkFIFO;
```

There are three rules defined in the module:

```
(* descending_urgency="enqueue_item, enqueue_bubble" *)
rule enqueue_item;
    outfifo.enq(infifo.first);
    infifo.deq;
    bubble_cycles <= 0;
endrule

rule inc_bubble_cycles;
    bubble_cycles <= bubble_cycles + 1;
endrule

rule enqueue_bubble;
    outfifo.enq(bubble_value);
    max_bubble_cycles <= max(max_bubble_cycles, bubble_cycles);
endrule
```

The rules enqueue something on the `outfifo` whenever it isn't full. They enqueue an item from the `infifo` whenever one is available (first rule), else they enqueue a "bubble_value" (third rule). Meanwhile, the rules also measure `max_bubble_cycles`, the longest continuous stretch of consecutive bubbles.

With the `descending_urgency` attribute, the user has specified that the rule `enqueue_item` is more urgent than the rule `enqueue_bubble`, that is, if it is possible to enqueue an item, it is more urgent to do that, than enqueue a bubble.

However, note that:

```
rule enqueue_item          writes          register bubble_cycles
rule inc_bubble_cycles     reads & writes  register bubble_cycles
rule enqueue_bubble        reads           register bubble_cycles
```

By normal ordering of register read and write methods, this implies that the *earliness* ordering (TRS ordering) of these rules must be:

```
enqueue_bubble, inc_bubble_cycles, enqueue_item
```

Thus, observe that in this example the urgency ordering is different from the earliness ordering.

Compile the example using the workstation. Verify that the flag -sched-dot is specified in the **Compile options** field on the **Compile** tab (in the **Project Options** window). This flag tells the compiler to generate .dot files containing graph files for each synthesized module.

Let's look at the schedule in the workstation. We're going to view the *Execution Order* graph (a.k.a. earliness) in the *Schedule Analysis* window.

1. Compile the example

2. Open the *Schedule Analysis* window (**Window→Schedule Analysis**)

3. Load the module mkGadget (**Module→Load**)

4. Display the *Execution Order* graph (**Scheduling Graphs→Execution Order**)[2]

As you can see in the graph in Figure 7.1, the execution order of the rules is:

1. enqueue_bubble

2. inc_bubble_cycles

3. enqueue_item

Specifically, note that the earliness ordering of enqueue_bubble and enqueue_item is opposed to their urgency ordering.

In summary:

- *Urgency* is a user-specified priority indicating in what order of importance rule must be considered for firing. This concerns rule conditions only, not rule bodies.

- *Earliness* is the logical sequencing of rules within a clock that is induced by the scheduling constraints on the methods they use (such as a read from a register must be earlier than a write to that register). This concerns both rule conditions and rule bodies.

---

[2]The package graphviz, including the Tcl extensions must be installed to view the scheduling graphs in the workstation. To view the scheduling graphs without the workstation you can use any 3rd-party package capable of displaying .dot files.

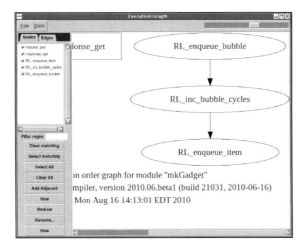

Figure 7.1: Execution Order Graph

7.2.3 mutually_exclusive

Source directory: `schedule/mutually_exclusive`

Source listing: Appendix A.6.4

The compiler's scheduling phase always attempts to deduce when two rules are mutually exclusive, i.e., prove that their conditions cannot simultaneously be true. However, the compiler may be unable to prove this, either because the scheduler effort limit is exceeded or because the mutual exclusion depends on information the scheduler does not know. The attribute `mutually_exclusive` allows the designer to assert to the scheduler that two rule conditions are mutually exclusive. The attribute `mutually_exclusive` may occur in the same syntactic positions as the attribute `descending_urgency`, and takes a similar argument, a string containing a comma-separated list of rule names. Example:

```
(* mutually_exclusive = "r1, r2, r3" *)
```

This asserts that every pair of rules in the list are mutually exclusive.

The `mutually_exclusive` assertion is sometimes called a "trust me" assertion, because the user is claiming a property that the compiler may be unable to prove on its own. However, the BSV compiler will generate code that will check and generate a runtime error if the two rules are ever enabled simultaneously in the same clock cycle during simulation. This allows you to find out when your use of the `mutually_exclusive` attribute is incorrect.

Let's look at an example. We have a 3-bit register named `x` which always has one bit set.

```
Reg#(Bit#(3)) x <- mkReg (1);
```

In rule rx we rotate the bits:

```
rule rx;
   cycle <= cycle + 1;
   x <= { x[1:0], x[2] };  // rotate the bits
   $display ("%0d: Rule rx", cycle);
endrule
```

Hence, x always has exactly one bit set, bit 0, or 1, or 2. Thus the rules, ry0, ry1, and ry2 always have mutually exclusive conditions. But this proof is too difficult for the compiler.

```
rule ry0 (x[0] == 1);
   y <= y + 1;
   $display ("%0d: Rule ry0", cycle);
endrule

rule ry1 (x[1] == 1);
   y <= y + 2;
   $display ("%0d: Rule ry1", cycle);
endrule

rule ry2 (x[2] == 1);
   y <= y + 3;
   $display ("%0d: Rule ry2", cycle);
endrule
```

Because the compiler does not know that the rules ry0, ry1 and ry2 are mutually exclusive, and each rule reads and writes register y, the compiler assumes a conflict between each pair of rules. Thus, it reports this during compilation and generates generates priority logic that suppresses the other rules when one of them is enabled.

With the mutually_exclusive attribute, however, the compiler assumes mutual exclusion of the rule conditions and does not generate priority logic.

When you compile the program *without* the mutually_exclusive attribute, the compiler generates the following warning messages, along with priority logic in the Verilog:

```
Warning: "Tb.bsv", line 15, column 8: (G0010)
  Rule "ry0" was treated as more urgent than "ry1". Conflicts:
    "ry0" cannot fire before "ry1": calls to y.write vs. y.read
    "ry1" cannot fire before "ry0": calls to y.write vs. y.read
Warning: "Tb.bsv", line 15, column 8: (G0010)
  Rule "ry1" was treated as more urgent than "ry2". Conflicts:
    "ry1" cannot fire before "ry2": calls to y.write vs. y.read
    "ry2" cannot fire before "ry1": calls to y.write vs. y.read
```

If you compile *with* the `mutually_exclusive` attribute, there is no conflict warning message and no priority logic generated.

Note: we get the same functional behavior with or without the assertion, since just one of those three rules is enabled at a time anyway. The difference is just in the quality of the Verilog—one has priority logic (more expensive) and the other does not.

7.2.4   conflict_free

Source directory: `schedule/conflict_free`

Source listing: Appendix A.6.5

Like the `mutually_exclusive` rule attribute, the `conflict_free` rule attribute is a way to overrule the scheduler's deduction about the relationship between two rules. However, unlike rules that are annotated `mutually_exclusive`, rules that are `conflict_free` may fire in the same clock cycle. The `conflict_free` attribute asserts that the annotated rules will not make method calls that are inconsistent with the generated schedule when they execute.

The `conflict_free` attribute may occur in the same syntactic positions as the `descending_urgency` attribute and takes a similar argument, a string containing a comma-separated list of rule names. Example:

```
(* conflict_free = "r1, r2, r3" *)
```

This asserts that every pair of rules in the list are conflict free.

For example, two rules may both conditionally enqueue data into a FIFO with a single enqueue port. Ordinarily, the scheduler would conclude that the two rules conflict since they are competing for a single method. However, if they are annotated as `conflict_free` the designer is asserting that when one rule is enqueuing into the FIFO, the other will not be, so the conflict is not real. With the annotation, the schedule will be generated as if no conflicts exist, and code will be inserted into the resulting model to check if conflicting methods are actually called by the conflict-free rules during simulation.

It is important to understand the `conflict_free` attribute's capabilities and limitations. The attribute works with more than method calls that totally conflict (like the single enqueue port). During simulation, it will check and report any method calls amongst `conflict_free` rules that are inconsistent with the generated schedule (including registers being read after they have been written and wires being written after they are read). On the other hand, the `conflict_free` attribute does not overrule the scheduler's deductions with respect to resource usage (like uses of a multi-ported register file).

Let's consider an example similar to the `mutually_exclusive` example in the previous section. In this case the rules `ry0`, `ry1`, and `ry2` are enabled on every cycle, but they access the shared resource y on mutually exclusive conditions.

```
    rule ry0;
        if (x[0] == 1) y <= y + 1;
    endrule

    rule ry1;
        if (x[1] == 1) y <= y + 2;
    endrule

    rule ry2;
        if (x[2] == 1) y <= y + 3;
    endrule
```

There is no conflict in the access to y, but the compiler does not know that. Thus it reports the conflict during compilation and generates priority logic that suppresses the other rules when one of them is enabled.

With the `conflict_free` attribute, the compiler assumes that the conflicting resource will never be accessed simultaneously, and does not generate priority logic. It does generate simulation code to ensure that the resource is never actually accessed simultaneously.

When you compile the example *without* the attribute, the compiler generates the following warning message along with priority logic:

```
Warning: "Tb.bsv", line 15, column 8: (G0010)
  Rule "ry0" was treated as more urgent than "ry1". Conflicts:
    "ry0" cannot fire before "ry1": calls to y.write vs. y.read
    "ry1" cannot fire before "ry0": calls to y.write vs. y.read
Warning: "Tb.bsv", line 15, column 8: (G0010)
  Rule "ry1" was treated as more urgent than "ry2". Conflicts:
    "ry1" cannot fire before "ry2": calls to y.write vs. y.read
    "ry2" cannot fire before "ry1": calls to y.write vs. y.read
Warning: "Tb.bsv", line 38, column 9: (G0021)
  According to the generated schedule, rule "ry1" can never fire.
Warning: "Tb.bsv", line 43, column 9: (G0021)
  According to the generated schedule, rule "ry2" can never fire.
```

If you compile *with* the `conflict_free` attribute, there are no conflicts warning messages and no priority logic generated.

7.2.5   preempts

Source directory: `schedule/preempts`

Source listing: Appendix A.6.6

The designer can also prevent a rule (or a set of rules) from firing whenever another rule fires. The `preempts` attribute accepts two elements as arguments. Each element must either be a rule name or a list of rule names. A list of rule names is enclosed in parentheses and separated by commas). In each cycle, if any of the rule names specified in the first list can be executed and are scheduled to fire, then none of the rules specified in the second list will be allowed to fire.

The `preempts` attribute is similar to the `descending_urgency` attribute and may occur in the same syntactic positions. The `preempts` attribute is equivalent to forcing a conflict and adding `descending_urgency`. With `descending_urgency`, if two rules do not conflict, then can still fire together. With `preempts`, since the first rule considered preempts the other if it fires, they can never fire together. If `r1` preempts `r2`, the compiler forces a conflict and gives `r1` priority. If `r1` is able to fire, but is not scheduled to fire, then `r2` can still fire.

If `r1` will fire, `r2` will not:

```
(* preempts = "r1, r2" *)
```

If either `r1` or `r2` (or both) will fire, `r3` will not:

```
(* preempts = "(r1, r2), r3" *)
```

If the rule `r1` in the sub-module `the_bar` will fire, then neither `r2` nor `r3` will fire:

```
(* preempts = "(the_bar.r1, (r2, r3)" *)
```

In this example we have a module `mkGadget`:

```
module mkGadget (Server#(int,int));
   FIFO#(int) infifo <- mkFIFO;
   FIFO#(int) outfifo <- mkFIFO;

   Reg#(int) idle_cycles <- mkReg(0);
```

The rule `enqueue_item` transfers an item from `infifo` to `outfifo` whenever possible:

```
rule enqueue_item;
   outfifo.enq(infifo.first);
   infifo.deq;
endrule
```

The rule `count_idle_cycles` counts the idle cycles:

```
rule count_idle_cycles;
    idle_cycles <= idle_cycles + 1;
    $display ("Idle cycle %0d", idle_cycles + 1);
endrule
```

Note that the two rules `enqueue_item` and `count_idle_cycles` do not conflict in any way. Thus, left to itself, `count_idle_cycles` would count every cycle. However, the line:

```
(* preempts="enqueue_item, count_idle_cycles" *)
```

ensures that `count_idle_cycles` only fires when `enqueue_item` does not fire, and thus it only counts cycles when no item is transferred between the queues.

In fact, if we compile the example in the development workstation, and open the **Schedule Analysis** window, we can view information about the rules. Load the `mkGadget` module (**Module→Load**) and select the **Rule Order** tab. Select **RL_count_idle_cycles**, to see the following information about the rule:

```
RL_count_idle_cycles

Predicate
True

Methods called in predicate

Methods called in body
idle_cycles.read
idle_cycles.write

Blocking rules
RL_enqueue_item
```

7.2.6   fire_when_enabled, no_implicit_conditions

Source directory: `schedule/fire_when_enabled`

Source listing: Appendix A.6.7

The `fire_when_enabled` and `no_implicit_conditions` attributes immediately precede a rule (just before the `rule` keyword) and govern the rule. They can only be attached to rule definitions, not module definitions like the scheduling attributes discussed earlier in this section.

The `fire_when_enabled` attribute asserts that this rule must fire whenever its predicate and its implicit conditions are true, i.e., when the rule conditions are true, the attribute checks that there are no scheduling conflicts with other rules that will prevent it from firing.

A rule may be inhibited from firing even if all its implicit and explicit conditions are true because it conflicts with some other rule which has higher priority. Further, during program development, as we add rules incrementally, this situation may change. The `fire_when_enabled` attribute is a useful compiler-checked assertion that ensures that a rule is not inhibited by other rules. It essentially asserts that the CAN FIRE signal of a rule is the same as its WILL FIRE, i.e., that if it can fire, it won't be inhibited by conflicts with any other rule.

The `no_implicit_conditions` attribute asserts that the implicit conditions of all interface methods called within the rule must always be true; only the explicit rule predicate controls whether the rule can fire or not.

A rule may call many methods, but unless we examine the implementations of those methods, we could not be sure whether those methods can be disabled or not by implicit conditions. Further, at different times during the design, the modules behind those method calls may be substituted, i.e., the method implementations can change. The `no_implicit_conditions` attribute is a useful compile-checked assertion that the method-calls in a rule are never disabled by implicit conditions

Both attributes are statically verified by the compiler and the compiler will report an error if necessary.

This example demonstrates the `fire_when_enabled` and `no_implicit_condition` attributes. In this example we have a register x, and a Put#(int) interface named dut.

```
module mkTb (Empty);

   Reg#(int) cycle <- mkReg (0);
   Reg#(int) x     <- mkReg (0);

   Put#(int) dut <- mkDut ();
```

Rule f1 has the attribute `no_implicit_conditions`, which asserts that the rule cannot be prevented from firing because some method called in the rule has an implicit condition.

```
   (* fire_when_enabled, no_implicit_conditions *)
   rule r1 (pack (cycle)[2:0] != 7);
      dut.put (x);
      x <= x + 1;
   endrule

   rule r2;
      x <= x + 2;
   endrule
```

However, the rule calls `dut.put()` and we can see that in `mkDut`, the `put()` method has an implicit condition:

```
module mkDut (Put#(int));
   Reg#(int) y <- mkReg (0);

   method Action put (int z) if (y > 0);
      y <= z;
   endmethod
endmodule
```

Thus the assertion is false. When we compile, we get the following scheduling error message:

```
Error: "Tb.bsv", line 38, column 9: (G0005)
   The assertion 'no_implicit_conditions' failed for rule 'r1'
```

We can get beyond this by commenting out the implicit condition:

```
method Action put (int z); // if (y > 0);
```

and recompiling. Now we get the following error message:

```
Error: "Tb.bsv", line 38, column 9: (G0005)
   The assertion 'fire_when_enabled' failed for rule 'RL_r1'
   because it is blocked by rule
     RL_r2
   in the scheduler
     esposito: [RL_r2 -> [], RL_r1 -> [RL_r2], RL_count_cycles -> []]
```

Even if the rule r1's explicit condition ('pack(cycle)[2:0] != 7') were true, the rule is still prevented from firing because:

- It conflicts with rule r2 (both r1 and r2 read and write register x)

- The descending_urgency attribute gives priority to r2 over r1

Thus, the fire_when_enabled assertion is false and we get the error message.

If we change the rule priorities as follows:

```
(* descending_urgency = "r1, r2" *)
```

it will compile successfully, because r1 can indeed execute whenever its conditions are true.

7.3   always_ready, always_enabled

Source directory: `schedule/always_ready`

Source listing: Appendix A.6.8

The method attributes `always_enabled` and `always_ready` assert some control properties of the annotated methods, and also remove the ports that thereby become unnecessary. These attributes are applied to synthesized modules, methods, interfaces, and subinterfaces at the top level only. If the module is not synthesized, the attribute is ignored. The compiler verifies that the attributes are correctly applied.

The attribute `always_enabled` can be applied to `Action` and `ActionValue` methods (value methods don't have enable signals). It asserts that the environment will invoke these methods on every clock. Thus, the normal enable signal in the generated Verilog is not necessary, and will be omitted. The compiler verifies this assertion, i.e., it verifies that the rule or other method from which this method is invoked fires on every cycle. If not, it emits a compile-time error.

The attribute `always_ready` can be applied to any method. It asserts that the condition on the method is always true. Thus, the normal ready signal in the generated Verilog is not necessary, and will be omitted. The compiler verifies this assertion, i.e., it attempts to prove that the method's condition is true. If not, it emits a compile-time error.

Note that `always_enabled` implies `always_ready`, since a rule calling a method that is not ready cannot be enabled.

The current example implements a parameterized model of a pipelined, synchronous ROM with 3-cycle latency. It accepts an address and delivers data on every cycle. If an address arrives on cycle J, the data is delivered on cycle J+3. Note: there is no handshaking, no flow control; `addr` and `data` are transferred on every clock (hence the *sync* in the name). The source includes three files: `Tb.bsv`, `PipelinedSyncROM_model.bsv`, and `mem_init.datat`.

The module `mkPipelinedSyncROM_model` is parameterized, and so it cannot be synthesized into a separate Verilog module. The first few lines of the testbench package

```
import PipelinedSyncROM_model::*;

typedef UInt#(16) Addr;
typedef UInt#(24) Data;

   (* synthesize *)
   (* always_ready = "request, response" *)
   (* always_enabled = "request" *)
   module mkPipelinedSyncROM_model_specific (PipelinedSyncROM#(Addr,Data));
```

makes a specific instance of the module, fixing the parameters; this specific instance can be synthesized into a separate Verilog module. The attributes:

```
(* always_ready = "request, response" *)
(* always_enabled = "request" *)
```

specify that both `request` and `response` methods are always ready, and that the `request` Action method is always enabled. The compiler will check that these assertions are indeed true, and then, in the generated Verilog, omit the usual `READY` and `ENABLE` signals.

The model can be compiled and run with both Bluesim and Verilog sim. Generate Verilog and observe the file `mkPipelinedSyncROM_model_specific.v`. When you examine the module header, you will see that there are no `READY` or `ENABLE` signals in the port list, as shown below.

```
module mkPipelinedSyncROM_model_specific(CLK,
                                         RST_N,
                                         request_a,
                                         response);
```

After you compile and observe the results, remove the `always_ready` and `always_enabled` attributes, recompile to Verilog and observe that in the Verilog ports we now have the `READY` and `ENABLE` signals for the methods.

7.4   -aggressive-conditions

Source directory: `schedule/aggressive_conditions`

Source listing: Appendix A.6.9

When a rule contains an `if`-statement, the implicit conditions in the branches must be propagated to the rule predicate. This can be done conservatively, by simply placing the implicit conditions for every branch in the predicate. Or it can be done more aggressively, by linking each implicit condition with its associated branch condition. The flag `-aggressive-conditions` turns on this feature, directing the compiler to attempt to link the conditions with the associated branches.

Let's look at an example in which we have two FIFOs, `f0` and `f1`, each of depth 3.

```
FIFO#(int) f0 <- mkSizedFIFO (3);
FIFO#(int) f1 <- mkSizedFIFO (3);
```

Rule `rA` tries to enqueue the cycle count into FIFO `f0` on even cycles and into FIFO `f1` on odd cycles. Rule `rb` simply drains FIFO `f0`.

```
rule rA;
    if (pack (cycle)[0] == 0) begin
        f0.enq (cycle);
        $display ("%0d: Enqueuing %d into fifo f0", cycle, cycle);
    end
    else begin
        f1.enq (cycle);
        $display ("%0d: Enqueuing %d into fifo f1", cycle, cycle);
    end
endrule

rule rB;
    f0.deq ();
endrule
```

Since FIFO f1 is not drained, it will become full after 3 enqueues, after which the f1.enq() method will be disabled. So, clearly, rule rA cannot do anything on odd cycles. The question is, can rule rA fire on even cycles, since in those cases it is not trying to enqueue on f1. The answer is *no* if we do not use the compiler's -aggressive-conditions flag, and *yes* if we do.

Open the project in the development workstation and look at the compile options. Let's first compile without the -aggressive-conditions flag. After compiling, open the **Schedule Analysis** window and load the top module. Look the **Rule Order** tab for rule RL_rA. You'll see that the predicate for the rule is:

f1.i_notFull && f0.i_notFull

Thus, the rule will not fire whenever f1 is full, regardless of whether it wants to enqueue on f1 or not. Indeed, when we simulate the program, we see that the rule stops firing after the three enqueues on f1:

```
0: Enqueuing        0 into fifo f0
1: Enqueuing        1 into fifo f1
2: Enqueuing        2 into fifo f0
3: Enqueuing        3 into fifo f1
4: Enqueuing        4 into fifo f0
5: Enqueuing        5 into fifo f1
```

Now let's try it with the -aggressive-conditions flag. Open the **Project Options** window and go to the **Compile** tab. In the **Compile options** field, add the flag -aggressive-conditions. Clean, recompile, link, and simulate (this is a single option on the **Build** menu and the toolbar). Now looking at the **Rule Order** tab for RL_rA, we see the predicate for the rule is:

```
cycle[0]
? f1.i_notFull
: f0.i_notFull
```

You can see that the rule predicate is more *aggressive*, more fine-grained in checking whether the rule can fire or not. It only checks whether f1 is full if it needs to enqueue on f1 and only checks whether f0 is full if it needs to enqueue on f0.

Thus, the rule can continue to fire on even cycles, when it is not attempting to enqueue on f1. When we run the program, we see that the rule continues firing. Until cycle 5, it enqueues on every cycle, alternating three items on f0 and f1. After that, it only enqueues f0 on even cycles.

```
 0: Enqueuing          0 into fifo f0
 1: Enqueuing          1 into fifo f1
 2: Enqueuing          2 into fifo f0
 3: Enqueuing          3 into fifo f1
 4: Enqueuing          4 into fifo f0
 5: Enqueuing          5 into fifo f1
 6: Enqueuing          6 into fifo f0
 8: Enqueuing          8 into fifo f0
10: Enqueuing         10 into fifo f0
12: Enqueuing         12 into fifo f0
14: Enqueuing         14 into fifo f0
16: Enqueuing         16 into fifo f0
```

Why not always use `aggressive-conditions`? Propagating branch conditions to rule predicates can add logic to your design. Also, aggressive conditions can increase compile time because the compiler's analysis of rule conditions must now work with more complex boolean expressions. Nevertheless, it is nowadays common practice to use `aggressive-conditions` routinely, and only switch it *off* if necessary for the above reasons.

Note that using `aggressive-conditions` is equivalent to splitting the rule into two separate rules and lifting the if's condition into the rule condition:

```
    rule rA_T (pack (cycle)[0] == 0);
       f0.enq (cycle);
       $display ("%0d: Enqueuing %d into fifo f0", cycle, cycle);
    endrule

    rule rA_F; (! (pack (cycle)[0] == 0));
       f1.enq (cycle);
       $display ("%0d: Enqueuing %d into fifo f1", cycle, cycle);
    endrule
```

Sometimes, splitting a rule syntactically like this makes the code more readable.

7.5   Separating an ActionValue method

Source directory: `schedule/separate_av`

Source listing: Appendix A.6.10

An ActionValue method combines an Action method and returns a value, combining the two types of methods (Action and value) into a single method. In this example, we are going to look at the difference between combining an Action and returning a value in a single ActionValue method vs. separating the Action into an Action method and returning the value in a value method.

We'll see that splitting an ActionValue method into an Action method and a value method can change the atomicity properties. In particular, whereas the ActionValue method is necessarily atomic, the separate Action and value methods can now be called from separate rules, and those rules can now have other intervening rules.

Let's examine the following interface definition:

```
interface Foo_ifc;
    method ActionValue#(int) avx ();
    method int vy ();
    method Action ay ();
```

And we have two registers that we are examining, x and y:

```
Reg#(int) x <- mkReg (0);
Reg#(int) y <- mkReg (0);
```

In module mkFoo, the rules incx and incy are identical, on registers x and y respectively:

```
rule incx;
    x <= x + 1;
    $display ("%0d: rule incx; new x = %0d", cycle, x+1);
endrule

rule incy;
    y <= y + 1;
    $display ("%0d: rule incy; new y = %0d", cycle, y+1);
endrule
```

The three methods in the module perform the following actions:

- The method avx returns the old value of x and sets x to 42.

- The method vy just returns the old value of y.

- The method ay just sets y to 42.

```
method ActionValue#(int) avx ();
    $display ("%0d: method avx; setting x to 42; returning old x = %0d",
             cycle, x);
    x <= 42;
    return x;
endmethod

method int vy ();
    return y;
endmethod

method Action ay ();
    $display ("%0d: method ay; setting y to 42", cycle);
    y <= 42;
endmethod
```

The rules in the top-level module mkTb call each of the methods:

- rule rAVX calls the ActionValue method avx

- rule rAY calls the Action method ay

- rule rVY calls the value method vy

```
rule rAVX;
    let x <- foo.avx ();
    $display ("%0d: rule rAVX, x = %0d", cycle, x);
endrule

rule rAY;
    $display ("%0d: rule rAY", cycle);
    foo.ay ();
endrule

rule rVY;
    let y = foo.vy ();
    $display ("%0d: rule rVY, y = %0d", cycle, y);
endrule
```

When we compile and run, we see output like this:

```
0: rule rVY, y = 0
0: rule incy; new y = 1
0: rule rAY
```

```
0: method ay; setting y to 42
0: method avx; setting x to 42; returning old x = 0
0: rule rAVX, x = 0
1: rule rVY, y = 42
1: rule incy; new y = 43
1: rule rAY
1: method ay; setting y to 42
1: method avx; setting x to 42; returning old x = 42
1: rule rAVX, x = 42
```

Note that in each cycle, with respect to the split methods, the rules rVY, incy, and rule rAY execute, in that order. On the other hand, looking at the combined method, rule rAVX fires, but not rule incx.

The explanation has to do with atomicity. The combined method avx both reads and writes register x, as does rule incx. Thus, the combined method conflicts with the rule and these two can never execute simultaneously in the same cycle. We can see which rules have blocking rules, and what they are on the **Rule Order** tab of the **Schedule Analysis** window.

On the other hand, in the split method, method vy only reads y; rule incy both reads and writes y, and method ay only writes y. Thus it is possible to schedule these in that order: vy before incy before ay. We see this in the **Execution Order** graph from the **Schedule Analysis** window, as shown in Figure 7.2.

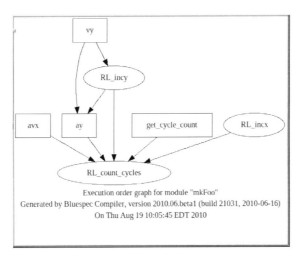

Figure 7.2: Execution Order Graph

Chapter 8

RWires and Wire types

A Wire is a primitive stateless module whose purpose is to allow data transfer between methods and rules without the cycle latency of a register. That is, a value can be written and read out in the same cycle; the value is not stored across clock cycles. Wires are used to pass data between rules within the same clock cycle. Wires in BSV are integrated into rule semantics, just like registers—wires are modules with interface methods, and the interface methods have schedules.

Before reading this section, it is important to be clear about the logical execution order of rules, as described in 5.2, and the BSV compiler's scheduling principle of firing a rule at most once within a clock cycle, as described in 7.

All wire types in BSV are built using the basic RWire type. But before we discuss the differences between RWire and its derived types (Wire, DWire, PulseWire, BypassWire), we need to understand the basic features of RWires, and hence, all wires.

- Wires truly become wires in hardware; they only carry values within a clock cycle, and they do not hold any *state* between cycles.

- A wire can be written at most once in a cycle.

- A wire's schedule requires that in a given cycle it is written before it is read, as opposed to a register that is read before it is written:
  Wire: write before read
  Register: read before write
  That is, any rule which writes to a wire must be scheduled earlier than any other rule which reads from the wire.

8.1 RWire

The RWire interface is defined (in the BSV library) as follows:

107

```
interface RWire#(type element_type) ;
   method Action               wset (element_type datain) ;
   method Maybe#(element_type) wget;
endinterface: RWire
```

The `wset` method writes to an RWire. The `wget` method reads the RWire, but instead of just returning a raw value, it is accompanied by a 'valid' bit. This combination of a value with a valid bit is a common paradigm in hardware design, and in BSV it is typically expressed using a `Maybe` type, which discussed in more detail in Section 10.1.5.

In a given cycle, if `wset`(v) is being invoked by a rule A, then `wget` returns, to its invoking rule B, "valid" along with v. If no rule A is currently invoking `wset`, then `wget` returns "invalid" to rule B. Thus,

- By testing the valid bit, rule B can "know" whether or not there is a rule A driving the RWire, and

- if so, rule B can also receive the value v communicated by rule A.

The valid test can be performed using the library function `isValid`(*mv*). The value carried on an RWire *mv* can be extracted using the function `fromMaybe`(*default_value, mv*). If *mv* is valid, the function returns the associated value carried by the Maybe; if it is invalid, it returns *default_value*.

In the example below, we supply ? as the default value, which represents a *don't care* value.

```
Maybe#(int) mbi = rw_incr.wget();
int    di       = fromMaybe (?, mbi);
```

The methods of an RWire are always ready. Note that if there is no rule B performing the `wget` while a rule A is performing a `wset`(v), then the communicated value v is "dropped on the floor", i.e., it is not observed by any rule.

8.1.1 Up-down counter with simultaneous up/down commands

Source directory: `rwire/simplerwire`

Source listing: Appendix A.7.1

Through the next few sections we are going to examine an up-down counter that can have simultaneous up and down commands (Action methods). Each example implements two different versions of the counter where the second version uses a form of Wire. In this first example, we use an RWire.

First we instantiate two up-down counters. They both provide the same interface type, `Counter`.

```
interface Counter;
    method int read();                    // Read the counter's value
    method Action increment (int di);     // Step the counter up by di
    method Action decrement (int dd);     // Step the counter down by dd
endinterface: Counter

module mkTb (Empty);
    Counter c1 <- mkCounter_v1;    // Instantiate a version 1 counter
    Counter c2 <- mkCounter_v2;    // Instantiate a version 2 counter
```

They are called from the testbench, mkTb, in identical ways from apparently identical rules. For both counters, the rules increment by 5 in states 0..6, decrement by 2 in states 4..10. The interesting states are 4, 5 and 6 where both increment and decrement rules are potentially attempted.

```
    rule incr1 (state < 7);   //increment counter1
        c1.increment (5);
    endrule

    rule decr1 (state > 3);   //decrement counter1
        c1.decrement (2);
    endrule

    rule incr2 (state < 7);   //increment counter2
        c2.increment (5);
    endrule

    rule decr2 (state > 3);   //decrement counter2
        c2.decrement (2);
    endrule
```

Version 1 of the counter is straightforward. It has no independently running rules. The increment and decrement methods perform their required actions directly. But note that, because they both read and write value1, the methods conflict. They cannot both execute in the same clock. Specifically, the rules incr1 and decr1 in mkTb cannot both fire in the same clock.

```
module mkCounter_v1 (Counter);    // Version 1 of the counter
   ...
   method Action increment (int di);
      value1 <= value1 + di;
   endmethod

   method Action decrement (int dd);
      value1 <= value1 - dd;
   endmethod
endmodule: mkCounter_v1
```

Version 2 of the counter uses RWires:

```
module mkCounter_v2 (Counter);
   Reg#(int) value2 <- mkReg(0);  // holding the counter's value

   RWire#(int) rw_incr <- mkRWire();  // Signal invoking increment
   RWire#(int) rw_decr <- mkRWire();  // Signal invoking decrement
```

These two RWires communicate from the `increment` and `decrement` methods to the `doit` rule. Both of them, if driven, carry an `int` value. We will discuss the `doit` rule after discussing the interface methods.

Whenever the `increment` method is invoked, it drives the RWire `rw_incr` carrying the value `di`. Specifically, whenever the rule that invokes `increment` fires (incr2), the RWire is driven.

```
      method Action increment (int di);
         rw_incr.wset (di);
      endmethod
```

Similarly, whenever the `decrement` method is invoked (from rule `decr2`), it drives the RWire `rw_decr`.

```
      method Action decrement (int dd);
         rw_decr.wset (dd);
      endmethod
```

These two methods do not conflict because they are just driving different RWires. Hence, these two methods, `increment` and `decrement`, can be invoked in the same clock cycle, and the rules `incr2` and `decr2` in `mkTB` can fire in the same clock cycle.

The rule `doit` below does all the work. Each `.wget()` method reads an RWire.

```
    (* fire_when_enabled, no_implicit_conditions *)
    rule doit;
       Maybe#(int) mbi = rw_incr.wget();
       Maybe#(int) mbd = rw_decr.wget();
       int   di        = fromMaybe (?, mbi);
       int   dd        = fromMaybe (?, mbd);
```

In the current clock, if an RWire is not being driven (by .wset, then .wget() returns Invalid. If the RWire is being driven with a value v, the .wget() returns Valid along with v.

The if-then-else structure tests all four combinations of whether the two RWire-reads are Invalid or Valid, that is, it test whether the methods increment and decrement are being invoked in the current clock cycle (neither, one of them, or both) and performs the appropriate action.

```
       if      ((! isValid (mbi)) && (! isValid (mbd)))
          noAction;
       else if (   isValid (mbi)  && (! isValid (mbd)))
          value2 <= value2 + di;
       else if ((! isValid (mbi)) &&    isValid (mbd))
          value2 <= value2 - dd;
       else    (   isValid (mbi)  &&    isValid (mbd))
          value2 <= value2 + di - dd;
    endrule
```

noAction is a special null Action (skip, do nothing). Thus, the rule doit performs the appropriate updates to the counter depending on which methods are being invoked.

8.1.2   Attributes (*fire_when_enabled, no_implicit_conditions*)

You may have noticed the attributes fire_when_enabled and no_implicit_conditions on the rule doit. These are intended to further guarantee safety. Even though the program will work without these attributes, we are relying on certain scheduling properties of the rule, and it is better to make this explicit so that:

- the compiler will ensure the scheduling properties are as expected

- it is obvious to anyone reading the program what the scheduling intentions are

For example, what will happen if the increment method is invoked in a particular cycle, and somehow the rule doit doesn't fire? The increment command and value will be missed (the value is "dropped on the floor"). To guarantee that this does not happen, we need to ensure that the rule always fires.

The `fire_when_enabled` attribute asserts that whenever the rule's conditions are true, the rule fires. A rule's condition is normally necessary, but not sufficient, for the rule to fire, because conflicts with other rules may prevent it from firing even when its condition is true. This assertion strengthens it, saying that the rule's conditions are indeed sufficient. This assertion is proved by the compiler, raising a compile-time error if false.

The `no_implicit_conditions` attribute asserts that none of the methods invoked in the rule has a condition that could be false. Since this rule also has no explicit conditions, this means that there are no firing conditions for the rule. This assertion is also proved by the compiler, raising a compile-time error if false.

Together these assertions ensure that this rule fires on every clock. The rule has no conditions, explicit or implicit, and is not prevented from firing due to conflicts with any other rule. This guarantees that neither the increment value nor the decrement value will be dropped on the floor.

8.1.3   Compiling the Example

When we compile, we see the following warning from the compiler:

```
Warning: "Tb.bsv", line 18, column 8: (G0010)
  Rule "decr1" was treated as more urgent than "incr1". Conflicts:
    "decr1" cannot fire before "incr1": calls to c1.decrement vs. c1.increment
    "incr1" cannot fire before "decr1": calls to c1.increment vs. c1.decrement
```

The compiler is stating that since `c1.increment` and `c1.decrement` conflict (they cannot be invoked in the same cycle), it cannot allow rules `incr1` and `incr2` to fire in the same cycle. In the absence of any other guidance, it picks `decr1` over `incr1`. Thus in cycles 0-3, when only incr1 can fire, it will fire. In cycles 4-6 when both rules are enabled, it will only allow `decr1` to fire; in cycles 7-10 when only `decr1` can fire, it will fire.

In the second counter, c2, there is no such warning about rules `incr2` and `decr2` because, as we have seen, they can fire together. So for counter c2, in cycles 4-6, both rules `incr2` and `decr2` will fire.

When we execute the code, the displayed outputs indeed show this behavior.

	Counter 1				Counter 2			
	Enabled		Fire		Enabled		Fire	
Cycles	incr1	decr1	incr1	decr1	incr1	decr1	incr1	decr1
0-3	✓		✓		✓		✓	
4-6	✓	✓		✓	✓	✓	✓	✓
7-10		✓		✓		✓		✓

8.1.4 RWire with pattern-matching

Source directory: `rwire/pattern_matching`

Source listing: Appendix A.7.2

Most of this example is identical to the previous one, except for the style in which we have written the body of the rule `doit`. This version uses a `case` statement with pattern-matching instead of the `if/then/else` statement. Pattern-matching is discussed in more detail in Section 10.1.6.

Instead of this:

```
     (* fire_when_enabled, no_implicit_conditions *)
     rule doit;
        Maybe#(int) mbi = rw_incr.wget();
        Maybe#(int) mbd = rw_decr.wget();
        int    di        = fromMaybe (?, mbi);
        int    dd        = fromMaybe (?, mbd);
        if     ((! isValid (mbi)) && (! isValid (mbd)))
           noAction;
        else if (   isValid (mbi)  && (! isValid (mbd)))
           value2 <= value2 + di;
        else if ((! isValid (mbi)) &&    isValid (mbd))
           value2 <= value2 - dd;
        else // (   isValid (mbi)  &&    isValid (mbd))
           value2 <= value2 + di - dd;
     endrule
```

we now write this:

```
     (* fire_when_enabled, no_implicit_conditions *)
     rule doit;
        case (tuple2 (rw_incr.wget(), rw_decr.wget())) matches
           { tagged Invalid,    tagged Invalid   } : noAction;
           { tagged Valid .di, tagged Invalid   } : value <= value + di;
           { tagged Invalid,    tagged Valid .dd } : value <= value - dd;
           { tagged Valid .di, tagged Valid .dd } : value <= value + di - dd;
        endcase
     endrule
```

The expression `tuple2(..., ...)` evaluates to a *two-tuple* or a *pair*. Think of it as a `struct` with two fields that are filled in with the values of the two expressions. In this case, the `tuple` contains the values of the RWires: `rw_incr.wget()` and `rw_decr.wget()`.

The `case(e) matches...endcase` construct is called a *pattern-matching* case (because of the presence of the keyword `matches`). It selects the first case that *matches the pattern* of the expression `e`.

In this example, the four lines enumerate the four combinations of whether the two RWire reads are Invalid or Valid, that is whether the methods `increment` and `decrement` are being invoked in the current clock cycle (neither, increment, decrement, or both). 2-tuples are matched with patterns like {pat2, pat2}, where `pat1` and `pat2` are themselves patterns matching the first and second components of the 2-tuple, respectively. A pattern like `tagged Valid .di` will only match a valid RWire value and allows us to use the name `di` in the arm of the case statement. Thus, for example, the second line of the case statement can be read as follows:

If the value of `rw_incr.wget()` is valid and carries a value `di` (which must be of type `int`), and the value of `rw_decr.wget` is invalid, then increment `value` by `di`.

Note that with this pattern-matching notation, when the value is `Invalid`, it is impossible to accidentally read the remaining 32 bits (which are garbage). Thus, pattern-matching notation not only enhances readability (and hence maintainability), but also safety (correctness).

When we execute the code, the displayed outputs show the same behavior as the previous example.

8.2   Wire

Source directory: `rwire/wire`

Source listing: Appendix A.7.3

Let's take the previous example and rewrite module `mkCounter_v2` to use the `Wire#()` data type instead of `RWire#()`.

`Wire` is an `RWire` that is designed to be interchangeable with a `Reg`. The `Wire` interface is a synonym for the `Reg` interface; you can replace a `Reg` with a `Wire` without changing the syntax.

```
typedef Reg#(type element_type) Wire#(type element_type);
```

The `Wire#()` interface and module are similar to `RWire#()`, but the valid bit is hidden from the user and the validity of the read is considered an implicit condition. The `Wire` interface works like the `Reg` interface, so mentioning the name of the wire gets (reads) its contents whenever they're valid, and using "<=" writes the wire. So instead of writing a value using `wset()`, we use the same notation as a register write, namely the non-blocking assignment operator <=. Instead of reading a `Maybe#()` value, we use the same notation as a register read; either we just mention the name of the Wire to read it implicitly or we can use the method `_read` explicitly.

The key point is that the information carried on the `Valid` bit of a `Maybe#()` type in an `RWire` is instead carried here as an implicit condition of the `_read()` method. That is, the `_read()` method is only enabled if the `_write()` (assignment) is happening earlier in the same clock cycle.

We declare our Wires as follows:

```
        // Signal that increment method is being invoked
        Wire#(int) w_incr <- mkWire();
        // Signal that decrement method is being invoked
        Wire#(int) w_decr <- mkWire();
```

The `increment()` and `decrement()` methods are changed to use register-write notation:

```
        method Action increment (int di);
            w_incr <= di;
        endmethod

        method Action decrement (int dd);
            w_decr <= dd;
        endmethod
```

The following rules capture the desired behavior:

```
        (* descending_urgency = "r_incr_and_decr, r_incr_only, r_decr_only" *)

        rule r_incr_and_decr;
            value2 <= value2 + w_incr - w_decr;
        endrule

        rule r_incr_only;
            value2 <= value2 + w_incr;
        endrule

        rule r_decr_only;
            value2 <= value2 - w_decr;
        endrule
```

Because of the implicit conditions on `w_incr` and `w_decr`, the first rule can only fire if both Wires are being assigned. Similarly, the second rule can only fire if `w_incr` is being assigned, and the third rule can only fire if `w_decr` is being assigned.

Note that all three rules are enabled if both `w_incr` and `w_decr` are being assigned. The attribute `descending_urgency` ensures that in this case the first rule is given priority over the other two.

When we execute the code, the displayed outputs indeed show the same behavior as the previous two examples.

8.3 DWire

Source directory: `rwire/dwire`

Source listing: Appendix A.7.4

Now let's look at the counter example rewriting module `mkCounter_v2` to use a `mkDWire` constructor instead of `mkWire`.

The `mkDWire` constructor also creates a module with a `Wire#()` interface (which is a synonym for the `Reg#()` interface). However, when using a `DWire`, the `_read` method is always enabled, and therefore has no implicit conditions. When using the DWire construct, if the Wire is assigned earlier in the current clock cycle, the `_read` method returns the value assigned. However, if the Wire is not assigned, the `_read` method returns a default value, which is provided as a parameter to the `mkDWire()` module constructor.

We declare the wires as follows:

```
// Signal that increment method is being invoked
Wire#(int) w_incr <- mkDWire (0);
// Signal that decrement method is being invoked
Wire#(int) w_decr <- mkDWire (0);
```

Notice that the default value on both wires is 0. Now we need only one rule to capture the desired behavior:

```
(* fire_when_enabled, no_implicit_conditions *)
rule r_incr_and_decr;
    value2 <= value2 + w_incr - w_decr;
endrule
```

This rule will fire on every clock (the `fire_when_enabled` and `no_implicit_conditions` attributes guarantee that). Whenever the `increment()` method is invoked, `w_incr` will carry the increment value, else it will carry 0. Whenever the `decrement()` method is invoked, it will carry the decrement value, else it will carry 0. Thus, on every clock, we add and subtract the appropriate value from `value2`.

Once again, when we execute the code, the displayed outputs show the same behavior as the previous examples.

8.4   PulseWire

Source directory: `rwire/pulsewire`

Source listing: Appendix A.7.5

In this variant of the counter example, the decrements are always by 1 instead of an arbitrary integer. Now the `decrement` method has no argument:

```
        rule decr1 (state > 3);
            // Decrement always by 1, so method has no argument
            c1.decrement ();
        endrule
    interface Counter;
        ...
        // Step the counter down by 1
        method Action decrement ();
    endinterface: Counter
```

In module `mkCounter_v2` in the earlier RWire example (Section 8.1.4), we used an RWire `rw_decr` to carry a value from the `decrement` method to the `doit` rule. We could continue to do this, redefining the `decrement` method to always pass the int (32 bit wide) constant 1, instead of an argument:

```
    method Action decrement ();
        rw_decr.wset (1);
    endmethod
```

Alternatively, we could modify the RWire declaration so that it carries a value that is *zero bits* wide:

```
    module mkCounter_v2 (Counter);
        ...
        RWire#(Int#(0)) rw_decr <- mkRWire();
```

Then, in the `decrement` method we use a don't care value for this 0-bit integer argument of `wset`:

```
    method Action decrement ();
        rw_decr.wset (?);    // 'Don't care' value for the wset argument
    endmethod
```

And in the pattern-match in rule `doit`, we use a *wildcard* symbol * in the patterns to match this value [10]:

```
    rule doit;
        case (tuple2 (rw_incr.wget(), rw_decr.wget())) matches
            { tagged Invalid,   tagged Invalid   } : noAction;
            { tagged Valid .di, tagged Invalid   } : value2 <= value2 + di;
            { tagged Invalid,   tagged Valid .*  } : value2 <= value2 - 1;
            { tagged Valid .di, tagged Valid .*  } : value2 <= value2 + di - 1;
        endcase
    endrule
```

But all these solutions with RWire seem like overkill. Instead, we could use a PulseWire. A PulseWire is an RWire without any data. It is useful within rules and action methods to signal other methods or rules in the same clock cycle.

The `PulseWire` interface is defined as:

```
interface PulseWire;
   method Action send();
   method Bool _read();
endinterface
```

The `send()` method is essentially the RWire's `wset()`, except it has no argument. The Valid/Invalid status is returned as the `Bool` value of the `_read` method. Because the method is called `_read`, just like in the `Reg#()` interface, we automatically reuse BSV's syntactic shorthand where the `_read()` method call is implicitly inserted by the compiler; we don't have to say `_read` explicitly.

PulseWires directly express the idea that there is no value carried along with the Valid/Invalid bit; the only thing of importance here is the Valid/Invalid bit itself.

First, we declare the PulseWire:

```
module mkCounter_v2 (Counter);
   ...
   // Signal that decrement method is being invoked
   PulseWire    pw_decr <- mkPulseWire();
```

In the `decrement` method, we just *send* the signal:

```
method Action decrement ();
   pw_decr.send ();
endmethod
```

And in the `doit` rule we implicitly `_read()` the PulseWire and matching it against the constants True and False:

```
rule doit;
   case (tuple2 (rw_incr.wget(), pw_decr)) matches
      { tagged Invalid,   False } : noAction;
      { tagged Valid .di, False } : value2 <= value2 + di;
      { tagged Invalid,   True  } : value2 <= value2 - 1;
      { tagged Valid .di, True  } : value2 <= value2 + di - 1;
   endcase
endrule
```

8.5 BypassWire

Source directory: `rwire/bypasswire`

Source listing: Appendix A.7.6

The final type of Wire that we want to examine in called a BypassWire. A BypassWire is an implementation of the Wire interface where the `_write` method is `always_enabled`. The compiler will issue a warning if the method does not appear to be called every clock cycle. The advantage of this tradeoff is that the `_read` method of this interface does not carry any implicit conditions, so it can satisfy a `no_implicit_conditions` assertion or an `always_ready` method.

This example is identical to the DWire example in Section 8.3, except that in module `mkCounter_v2` we use the `mkBypassWire` constructor instead of the `mkDWire` constructor for the increment signal. We have also modified the testbench `mkTb` to invoke the `increment` method on every clock, instead of just on clocks 0 to 6.

In the module `mkCounter_v2`, we create the `w_incr` wire module to signal that increment method is being invoked:

```
Wire#(int) w_incr <- mkBypassWire();
```

We have a single rule that can always fire, because neither `w_incr` (ByPassWire) nor `w_decr` (DWire) have implicit conditions on their `_read` methods:

```
(* fire_when_enabled, no_implicit_conditions *)
rule r_incr_and_decr;
   value2 <= value2 + w_incr - w_decr;
endrule
```

The `increment` method:

```
method Action increment (int di);
    w_incr <= di;
endmethod
```

must now be called on every clock, since it is assigning to a BypassWire, which assumes the assignment is always enabled. The attribute:

```
always_enabled = "increment"
```

just above the `module mkCounter_v2` header asserts this property and is thus verified by the compiler.

Accordingly, in the testbench the `incr1` rule (and similarly the `incr2` rule) now invoke the `increment` method in all states:

```
    rule incr1;
        c1.increment (5);
    endrule
```

8.6   RWires and atomicity

Source directory: `rwire/rwire_atomicity`

Source listing: Appendix A.7.7

You must be careful when using RWires. When you split actions across multiple rules, you cannot assume that they are still atomic. This is of course true in any rule-splitting, whether involving RWires or not, but we discuss this topic here because RWires are typically used to split rules.

In this example we're going to look at a module `mkPipelineFIFO` which implements a 1-element FIFO. As with any FIFO, one can enqueue data when empty. When full, one can not only dequeue the old data, but in the same cycle one can also enqueue new data (hence the *pipeline* in the name because it behaves like an interlocked pipeline register).

As in the up/down counter examples, we can use RWires when two methods have to read and write common state (the *full* bit), if those methods are to run simultaneously in the same clock.

```
module mkPipelineFIFO (FIFO#(int));

    Reg#(Bool)        full      <- mkReg (False);
    Reg#(int)         data      <- mkRegU;

    RWire#(int)       rw_enq    <- mkRWire;              // enq method signal
    PulseWire         pw_deq    <- mkPulseWire;          // deq method signal

    Bool enq_ok = ((! full) || pw_deq);
    Bool deq_ok = full;
```

Each method sets an RWire and then a common rule (`rule_update_final`) does the actual work.

```
    rule rule_update_final (isValid(rw_enq.wget) || pw_deq);
        full <= isValid(rw_enq.wget);
        data <= fromMaybe (?, rw_enq.wget);
    endrule
```

Looking at the methods in `mkPipelineFIFO`, both `enq()` and `deq()` read the `full` register (as part of `enq_ok` and `deq_ok`). The `clear()` method writes the `full` register. So the `clear()` method must be scheduled *after* the `enq()` and the `deq()` methods.

```
    method Action enq(v) if (enq_ok);
       rw_enq.wset(v);
    endmethod

    method Action deq()  if (deq_ok);
       pw_deq.send ();
    endmethod

    method first()        if (full);
       return data;
    endmethod

    method Action clear();
       full <= False;
    endmethod
 endmodule: mkPipelineFIFO
 endpackage: Tb
```

In module mkTb, we instantiate such a FIFO and enqueue and dequeue some data on every cycle. Additionally, in cycle 2, we also invoke the clear method to forcibly empty the FIFO. In fact, in cycle 2, all four methods, enq(), first(), deq(), and clear(), are invoked. As we saw before, clear is scheduled last. In which case, we would expect that at the end of cycle 2, the FIFO is left empty (cleared).

```
module mkTb (Empty);
   ...
   rule e;
      let x = state * 10;
      f.enq (x);
   endrule

   rule d;
      let y = f.first ();
      f.deq ();
   endrule

   // Also clear the FIFO in state 2
   rule clear_counter (state == 2);
      f.clear ();
      $display ("State %0d: Clearing", state);
   endrule
```

But a transcript of the run shows:

State	Action
State 0	Enqueue 0
State 1	Dequeue 0
State 1	Enqueue 10
State 2	Dequeue 10
State 2	Enqueue 20
State 2	Clearing
State 3	Dequeue 20
State 3	Enqueue 30
...	...

In state 3 we successfully dequeue 20, which indicates that the FIFO did not get cleared at all. The value 20 actually got enqueued in state 2!

This looks like a bug. The schedule says `enq(20)` precedes `clear()`, but after this the FIFO contains 20 and is not cleared.

The resolution of this paradox is that we mistakenly imagined that the entire enqueue operation was atomic. However, the enqueue operation is now split into two different rules, one which calls the `enq()` method, and later the `rule_update_final` rule. So the *logical* enqueue operation is no longer a single atomic action. What happened here is that the `clear()` method got scheduled in between the those two rules. In fact, during compilation, we are told about this:

```
Warning: "Tb.bsv", line 64, column 8: (G0036)
  Rule "clear" will appear to fire before "rule_update_final" when both fire
  in the same clock cycle, affecting:
    calls to full.write vs. full.write
```

Thus, even though the `enq()` method preceded the `clear()` method, the final enqueue action came later, and we see its effects.

Bottom line: when you split actions across multiple rules communicating using RWires, you cannot assume they are atomic. Be careful when using RWires.

Chapter 9

Polymorphism

In BSV, extreme re-use is possible via polymorphic functions and modules. A polymorphic function is a function that can operate on more than one type of data. A classic example in higher level languages is to calculate the length of a list. The function can operate on many types of lists; the data in the list can be anything, char, int, float, complex, struct, etc. The number of items is returned regardless of the data types of the elements in the list.

We use polymorphic modules and functions often in BSV, because they allow us to isolate the problem we are trying to solve and then to reuse that code for as wide a selection of data types as possible. The BSV library packages define many polymorphic modules and functions, such as the `Client`, `Server` and `FIFO` interfaces in the previous example.

There may be restrictions on the data types that a given polymorphic module or function can work with. For instance, the addition operator + works on the data types for which addition makes sense: `Int`, `UInt`, and `Bit`, but not `Bool` or structures or other data types that don't explicitly define addition. The requirements of the type of data a module or function can work with are specified via *provisos*.

When a module or function is processed by the compiler, it generates a signature which contains its inputs, outputs, and valid data types. Provisos place constraints on the data types that can match that signature.

In the next few examples we'll create some polymorphic functions. Since modules are just a special kind of function, the same ideas apply directly to modules as well.

9.1   Polymorphic function

Source directory: `polymorphism/function`

Source listing: Appendix A.8.1

Part of the definition of a function is the type of the value returned by the function. We can use a variable for the name of type returned, instead of stating a specific type. This variable's name is used only for the function itself. Since it is a variable (even though the value of the variable is a type) it must start with a lower case letter. Often, we name it t*something*, the t indicating that it is representing a type variable.

The following function does nothing but pass its argument (of type td) through:

```
function td nop( td i ); return i; endfunction
```

When writing polymorphic functions, we often need to tell the compiler something about the data we are using. This restricts the types that can be used with the function. We use *provisos* to describe the types of data which the function can use.

In the example below, we are writing a function named equal to compare two items. In this case we have used the overloaded equality operator (==) on the data items. The items are of the same type, represented by the type variable td. Since we've used the equality function, the type represented by td must be in the Eq type class. We specify this requirement by adding the provisos Eq#(td). This says that only data types that have the == function defined can use our newly defined function.

```
function Bool equal( td i, td j )
    provisos( Eq#(td) );
  return (i == j);
endfunction
```

In this next function, add2x, we use the overloaded addition (+) operator, an arithmetic function defined in the **Arith** typeclass. Therefore td must be a type in the **Arith** typeclass. The other overloaded arithmetic operators (*, /, -) use the same **Arith** proviso, since they are also defined by the **Arith** typeclass. Specifying the proviso prevents the user from calling the function with an inapplicable data type, for example, a Bool which doesn't have + defined (it doesn't make sense to add two Bools).

```
function td add2x( td i, td j )
    provisos( Arith#(td) );
  return ( i + 2*j );
endfunction
```

Some functions require more than one proviso. In this next function, i2xj, we are using both arithmetic and equality operators, so we need both the **Arith** and Eq provisos.

```
function Bool i2xj( td i, td j )
    provisos( Arith#(td), Eq#(td) );
  return ( (i - 2*j) == 0 );
endfunction
```

Probably the most common proviso is `Bits#(type, sz)`. This proviso makes sure that the data type is in the `Bits` type class, ensuring that the overloaded functions `pack` and `unpack` are defined for the data type. This effectively allows you to perform any bit manipulation you require, because the data type has `pack` and `unpack` as a means of converting a type to bits and back. You must be careful doing this as you strip away the *real* type information.

In the next function, `isBlock`, we want to do a logical AND to see if a number is divisible by 4. The function uses `pack` to convert the data type to the bits data type, which has `&` and `==` defined. The `Bits` proviso ensures that `pack` is defined.

```
function Bool isBlock( td i )
      provisos(Bits#(td,szTD));
    return (pack(i) & 3) == 0;
endfunction
```

We also want to prevent anyone from using this function with a type that is less than 2 bits wide (otherwise the &3 won't make sense). We can use the size variable `szTD` generated by Bits. `szTD` is a local variable, valid only within the function. The proviso `Add#(x,y,z)` says that $x + y$ must equal z. The common trick is that the proviso `2+ unused = szTD` means that `szTD` >=2 since sizes must be greater than or equal to 0. We add this additional proviso to ensure that the data type is at least 2 bits wide.

```
function Bool isBlock2( td i )
      provisos(Bits#(td,szTD),
               Add#(2,unused,szTD));   // 2 + unused == szTD
    return (pack(i) & 3) == 0;
endfunction
```

This next function, `isMod3`, checks to see that the value passed in is divisible by 3. Since we aren't packing or unpacking, we don't need the `Bits` proviso. But we are specifying that the value should be greater than 4 bits. Use the `Add#()` with the `SizeOf#()` pseudo function to get the bit width of a type.

```
function Bool isMod3( td i )
       provisos(Arith#(td),
                Eq#(td),
                Add#(4,unused,SizeOf#(td)));
     return (i % 3) == 0;
endfunction
```

9.1.1   Common overloading provisos

The following table lists the commonly used type class provisos, along with the functions defined by the type class enforced by the proviso.

Common Overloading Provisos		
Proviso	Type class	Overloaded operators defined on type t
`Bits#(t,n)`	Bits	pack/unpack to convert to/from `Bit#(n)`
`Eq#(t)`	Eq	`==` and `!=`
`Arith#(t)`	Arith	`+`, `-`, `*`, `negate`, `/`, `%`
`Literal#(t)`	Literal	`fromInteger()` to convert an integer literal to type t. and `inLiteralRange` to determine if an Integer is in the range of the target type t
`Ord#(t)`	Ord	order-comparison operators `<.<=`, `>`, `>=`
`Bounded#(t)`	Bounded	`minBound` and `maxBound`
`Bitwise#(t)`	Bitwise	functions which compare two operands bit by bit to calculate a result
`BitReduction#(t)`	BitReduction	functions which take a sized type and reduce to one element.
`BitExtend#(t)`	BitExtend	extend, zeroExtend, signExtend, and truncate

9.1.2 Size relationship provisos

These classes are used in provisos to express constraints between the sizes of types.

Class	Proviso	Description
`Add`	`Add#(n1,n2,n3)`	Assert $n1 + n2 = n3$
`Mul`	`Mul#(n1,n2,n3)`	Assert $n1 * n2 = n3$
`Div`	`Div#(n1,n2,n3)`	Assert ceiling $n1/n2 = n3$
`Max`	`Max#(n1,n2,n3)`	Assert $\max(n1, n2) = n3$
`Min`	`Min#(n1,n2,n3)`	Assert $\min(n1, n2) = n3$
`Log`	`Log#(n1,n2)`	Assert ceiling $\log_2(n1) = n2$.

Examples of Provisos using size relationships:

```
    instance Bits #( Vector#(vsize, element_type), tsize)
       provisos (Bits#(element_type, sizea),
                 Mul#(vsize, sizea, tsize));        // vsize * sizea = tsize

    function Vector#(vsize1, element_type)
         cons (element_type elem, Vector#(vsize, element_type) vect)
       provisos (Add#(1, vsize, vsize1));          // 1 + vsize = vsize1

    function Vector#(mvsize,element_type)
          concat(Vector#(m,Vector#(n,element_type)) xss)
        provisos (Mul#(m,n,mvsize));               // m * n = mvsize
```

9.1.3 valueOf and SizeOf pseudo-functions

Bluespec provides these pseudo-functions to convert between types and numeric values. The pseudo-function `valueof` (or `valueOf`) is used to convert a numeric type into the corresponding Integer value.

The pseudo-function `SizeOf` is used to convert a type t into the numeric type representing its bit size.

valueof valueOf	Converts a numeric type into its Integer value.
	`function Integer valueOf (t) ;`

SizeOf	Converts a type into a numeric type representing its bit size.
	`function t SizeOf#(any_type)` `    provisos (Bits#(any_type, sa)) ;`

Example:

```
typedef Bit#(8) MyType;
//MyType is an alias of Bit#(8)

typedef SizeOf#(MyType) NumberOfBits;
//NumberOfBits is a numeric type, its value is 8

Integer ordinaryNumber = valueOf(NumberOfBits};
//valueOf converts a numeric type into an Integer
```

9.2 Simple polymorphic module with provisos

Source directory: `polymorphism/provisos`

Source listing: Appendix A.8.2

In this section we're going to look at a set of examples reusing the `Reg#(td)` interface. Often when using the `Reg#(td)` interface we need some provisos. The value stored in a `Reg` interface must always be in the `Bits` type class, that is the operations `pack` and `unpack` are defined on the type to convert into bits and back. So register interfaces always have the `Bits` proviso.

This first example is just hiding a register under a new name. We cannot synthesize this boundary because it is polymorphic. In order to be synthesizable, a module must meet the following characteristics:

- The module must be of type `Module` (advanced use of modules permits richer module types than normal modules);

- Its interface must be fully specified; there can be no polymorphic types in the interface;

- Its interface is a type whose methods and subinterfaces are all convertible to wires;

- All other inputs to the module must be convertible to Bits.

So to be able to synthesize a polymorphic module you have to wrap, or instantiate it from another module, with a specific data type.

First, let's look at an example, mkPlainReg, which provides a Reg#(tx) interface:

```
module mkPlainReg( Reg#(tx) ) provisos(Bits#(tx,szTX));
   Reg#(tx) val <- mkRegU;
   return val;
endmodule
```

Since the interface provided by mkPlainReg and the interface of val are both Reg#(tx), we can just return the interface as shown in the return val statement. Since mkRegU has the requirement of being in the Bits class, the proviso Bits#(tx, szTX) is required by mkPlainReg as well.

The above module provides any Reg#(tx) interface which meets the Bits requirement, as stated by the proviso. We could also define a register that returns Bit#(n):

```
module mkPlainReg2( Reg#(Bit#(n)) );
   Reg#(Bit#(n)) val <- mkRegU;
   return val;
endmodule
```

This is still polymorphic, because we haven't defined the width of the register. But the proviso is not required because we know the data type (we don't know the size), and by defining Bit#(n) we have pack and unpack defined.

Once we wrap our polymorphic module with a specific data type we can synthesize it. In the mkPlainRegInt module we instantiate the mkPlainReg module with a value of int. We are reusing the mkPlainReg module for the specific int data type.

```
(* synthesize *)
module mkPlainRegInt( Reg#(int) );
   Reg#(int) val <- mkPlainReg;

   method Action _write(int i) = val._write(i);
   method int    _read()       = val._read();
endmodule
```

In this next example, we are going to also need the `Bitwise#(tx)` proviso, because we are using the bitwise invert function. The module can only be instantiated with a data type that is in the Bitwise type class. The module could never be instantiated with a Bool, for instance.

```
module mkNotReg( Reg#(tx) )
    provisos(Bits#(tx,szTX), Bitwise#(tx) );
   Reg#(tx) val <- mkRegU;

   method Action _write(tx i) = val._write(i);

   method tx     _read();
      return ~val;  // Bitwise#(tx) added because of ~ here
   endmethod
endmodule
```

The `mkDoubleReg` module, defined below, uses an arithmetic operator, requiring the `Arith` proviso.

```
module mkDoubleReg( Reg#(tx) )
    provisos(Bits#(tx,szTX), Arith#(tx));
   Reg#(tx) val <- mkRegU;

   method Action _write(tx i) = val._write(i);

   method tx     _read();
      return val * 2; // Arith#(tx) added here
   endmethod
endmodule
```

As a general guideline, we recommend that you get a non-polymorphic version of the module working before you attempt to parameterize a module. One reason is because the error messages provided can be easier to understand when real data types are used.

Once you start parameterizing your blocks, you need to understand the error messages put out by the compiler. Let's comment out the `Bitwise` proviso in the `mkNotReg` example:

```
module mkNotReg( Reg#(tx) )
    provisos(Bits#(tx,szTX) );

  Reg#(tx) val <- mkRegU;

  method Action _write(tx i) = val._write(i);

  method tx      _read();
      return ~val;  // Bitwise#(tx) added because of ~ here
  endmethod
endmodule
```

When you compile, you will get this error message:

```
Error: "Tb.bsv", line 61, column 8: (T0030)
  The provisos for this expression are too general.
  Given type:
    _m__#(Reg#(tx))
  With the following provisos:
    IsModule#(_m__, _c__)
    Bits#(tx, szTX)
  The following additional provisos are needed:
    Bitwise#(tx)
    Introduced at the following locations:
      "Tb.bsv", line 69, column 14
```

The general error you will often see is T0030: "The provisos for this expression are too general." This means there are not enough provisos for the compiler to be able to safely match the module to the various data types when called. In this case, the compiler tells us which proviso is missing.

A frequent issue is that blocks may have state of differing sizes. In this example, we put the type on the interface for a module, since the size of the inputs and outputs affects the size of bus width going in and out of the blocks.

```
interface T2#(type type_a, type type_b);
   method Action      drive(type_a ina, type_b inb);
   method type_a      result();
endinterface
```

In the module definition below, type_a is the same as type_x and type_b is the same as type_y. We could use the same variable names, but we're using different names to demonstrate that they don't have to be the same name.

```
module mkT2( T2#(type_x,type_y) )
    provisos(Bits#(type_y,sY),  // need to pack/unpack type_y
             Arith#(type_x),    // need to add with type "type_a"
             Add#(SizeOf#(type_x),0,SizeOf#(type_y))); // must be the same size
```

Since `type_b` is a different type than `type_a` we need to *recast* it via pack/unpack. In this case it is the same width, which is determined by the Add provisos above.

```
    Wire#(type_x) val <- mkWire;

    method Action  drive(type_x ina, type_y inb);
        val <= ina + unpack(pack(inb));
    endmethod

    method type_x      result();
        return val;
    endmethod
endmodule
```

Now let's modify the above example to make them different sizes. For example, maybe we send in `type_a`, but return a wider size, or maybe perhaps, a log size. In this interface we send in `type_a` and return `type_b`:

```
interface T3#(type type_a, type type_b);
    method Action      drive(type_a ina);
    method type_b      result();
endinterface
```

Now we will have to add provisos to relate `type_a` and `type_b`. In the previous example, we used the Add proviso to define them as the same.

```
module mkT3( T3#(type_a,type_b) )
    provisos(Bits#(type_a,sA),    // need to pack/unpack type_a
             Bits#(type_b,sB),
             Add#(unused,sB,sA),  // needed by truncate
             Log#(sA,sB));        // just for fun, make output roof of log2
```

It can get confusing sorting out the packs and unpacks, so it's often easiest to break out the individual steps, rather than trying to do it all at once. This has the added advantage of allowing the compiler to give you more accurate error messages.

```
   Wire#(type_b) val <- mkWire;

   method Action  drive(type_a ina);

      Bit#(sA) tmp  = pack(ina);
      Bit#(sB) tmp2 = truncate(tmp);
      val           <= unpack(tmp2);

      // the above 3 lines could be rewritten as:
      // val <= unpack(truncate(pack(ina)));
   endmethod

   method type_b      result();
      return val;
   endmethod
endmodule
```

Chapter 10

Advanced types and pattern-matching

10.1 User-defined types

While BSV provides many pre-defined types, users can define their own types including enumerated types, structures, and tagged unions through the `typedef` statement. User-defined types can only be defined at the top level of a package.

10.1.1 Type synonyms

The simplest user-defined type is a type synonym which allows you to define shorter or more meaningful names for existing types. Type synonyms are just for convenience and readability. The new type and the original type are the same type with different names; they can be used interchangeably.

```
typedef Type Identifier;
```

Examples:

```
typedef Int#(32) TDataX;
typedef Int#(32) TDataY;
typedef Bool     TFlag;

TDataX a = 10;
TDataY b = a;    // this is ok, since TDataX and TDataY are
                 // synonyms of the same type
```

133

However, when using `typedef` to define a new enum, struct, or tagged union type, the new type is not a synonym for any existing type. These introduce new types that are different from all other types. For example, even if two `typedef`s give names to struct types with exactly the same corresponding member names and types, they define two distinct types ("name equality", not "structural equality"). This is another manifestation of the principle of strong type-checking.

10.1.2 Enums

Source directory: `adv_types/enum`

Source listing: Appendix A.9.1

An `enum` defines a new type consisting of a set of scalar values with symbolic names.

```
typedef enum {Identifier1, Identifier2, ... IdentifierN} EnumName
    deriving (Eq, Bits);
```

- Enum label identifiers must begin with uppercase letters.

- Enum labels can be repeated in different enum definitions (the type-checker will disambiguate such identifiers).

- The optional `deriving Bits` clause tells the compiler to generate a bit-representation for the enum, allowing values of the type to be used by all facilities and functions that require this (as a proviso), including the `Reg#()` and `FIFO#()` types.

  When `deriving Bits` is specified, the default decoding of labels is 0,1,2..., using just enough bits to encode the entire set.

- The optional `deriving Eq` clause tells the compiler to pick a default equality operation for the labels, allowing values of the type to be tested for equality and inequality.

Examples:

```
    typedef enum {Green, Yellow, Red} TrafficLight deriving (Eq, Bits);
    typedef enum {Reset, Count, Decision} States deriving (Eq, Bits);
    typedef enum {IDLE, DOX, DOY, FINISH} Tenum1 deriving (Bits, Eq);
```

The enum declaration may explicitly specify the encoding used by assigning numbers to the labels. When an assignment is omitted, the label is assigned the previous label value plus one. If the initial label is omitted, it defaults to zero. Encodings must be unique, i.e., you cannot specify the same encoding for more than one label. For the `Tenum1` example above, the default label values would be (0,1,2,3).

The labels are assigned explicitly in the following example.

```
typedef enum { S0=43, S1, S2=20, S3=254 } Tenum2 deriving(Bits, Eq);
```

Since S1 doesn't have a label specified, it is assigned the value 44.

When deriving Bits is specified, the compiler generates a Bit representation for the enum. The width of its values is just enough to hold the largest value. In the example Tenum1, the largest value is 2, requiring a Bit#(2) to represent the values. In the Tenum2 example, the largest value is 254, requiring a Bit#(8) to represent the values.

The current example shows some enum declarations, and then some registers to hold those enum values.

```
package Tb;

typedef enum { IDLE, DOX, DOY, FINISH } Tenum1 deriving(Bits, Eq);
typedef enum { S0=43, S1, S2=20, S3=254 } Tenum2 deriving(Bits, Eq);

(* synthesize *)
module mkTb (Empty);
   Reg#(int) step <- mkReg(0);

   // note r1 is initialized with one of the enum value
   Reg#(Tenum1) r1 <- mkReg( IDLE );

   // note r2 is initialized with one of the enum value
   Reg#(Tenum2) r2 <- mkReg( S3 );
```

You cannot do arithmetic on enums as they are values, not bits. This statement:

```
      Tenum1 foo = IDLE;
      foo = foo + 1;
```

will cause the following error from the compiler:

```
Error: "Tb.bsv", line 58, column 17: (T0031)
  The provisos for this expression could not be resolved because there are no
  instances of the form:
    Arith#(Tb::Tenum1)
```

Again, this is a manifestation of the principle of strong type-checking and representation independence. You can of course explicitly choose to define the arithmetic operators on a particular enum type by making it an instance of the Arith typeclass and specifying the meanings of the arithmetic operators.

10.1.3　Structs

A structure or record (struct) is a composite type made up of a collection of members. Each member has a particular type and is identified by a member, or field, name. A struct value contains all the members. This is different from a tagged union (Section 10.1.4), in which each value contains one of the members. Another way to think of structs and tagged unions is that a struct is a conjunction—member1 *and* member2 ... *and* memberN–whereas a tagged union is a disjunction—member1 *or* member2 ... *or* memberN.

- A struct value contains all members

- The name of the struct begins with an upper case letter

- Struct member names each begin with a lower case letter

- Each member name in a given structure must be unique within that struct, but may be used in other structs or tagged unions

Examples:

```
typedef struct { int x; int y; } Coord;
typedef struct { Addr pc; RegFile rf; Memory mem; }  Proc;

typedef struct {
   Bool      flag;
   TDataY    coordX;
   TDataX    coordY;
 } TName deriving(Bits, Eq);

typedef struct{
    Vector#(3, Bit#(8))    color;
    Bit#(4)                brightness;
    Bool                   blinking;
} Pixel
   deriving (Bits, Eq);
```

Note that structured types can be arbitrarily nested. In the `Pixel` example, the `color` field is itself a `Vector`. The `Pixel` type uses `deriving (Bits, Eq)` to instruct the compiler to define the default bit representation (shown in Figure 10.1) and equality operators this type.

To access a structure member, use the StructName.memberName notation. Example using the Pixel type:

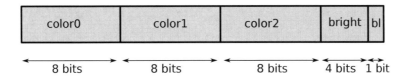

Figure 10.1: Pixel Structure

```
Vector#(256, Pixel)   screenLine;

screenLine[0].color[1]    = 8'hFF;
screenLine[0].brightness  = screenLine[0].brightness + 1;
screenLine[0].blinking    = False;
```

You can also write a "struct expression" to construct an entire struct value. Example (the right-hand side of the assignment):

```
let myPixel = Pixel { color: replicate (8'hFF),
                      brightness: 4'd2,
                      blinking: screenLine[3].blinking };
```

10.1.4   Tagged unions

A tagged union is a composite type made up of a collection of members. A union value only contains one member at a time while a structure value contains all of its members. Another way to think of structs and tagged unions is that a struct is a conjunction—member1 *and* member2 ... *and* memberN–whereas a tagged union is a disjunction—member1 *or* member2 ... *or* memberN.

Union member names begin with an uppercase letter. Each member name in a particular union must be unique within the union, but may be used in other structs or tagged unions. In a tagged union the member names are called *tags*. In the representation of a tagged union, in addition to the bits corresponding to the member types, a tagged union typically also has some bits to remember the tag. You can only access a single member at a time, corresponding to the current tag value.

Example:

```
typedef union tagged {
    Bit#(5)  Register;
    Bit#(22) Literal;
     struct {
         Bit#(5) regAddr;
         Bit#(5) regIndex;
     } Indexed;
} InstrOperand
   deriving (Bits, Eq);
```

An instance of `InstrOperand` is either a 5-bit register specifier *or* a 22-bit literal value *or* an indexed memory specifier consisting of two 5-bit register specifiers. This example also again demonstrates that structured types (structs, tagged unions, vectors, etc.) can be arbitrarily nested.

Because the type declaration uses `deriving (Bits, Eq)`, the compiler defines a canonical bit representation and equality operations for the type. The canonical bit representation is shown in Figure 10.2. The canonical representation takes the canonical representation of each of the members and

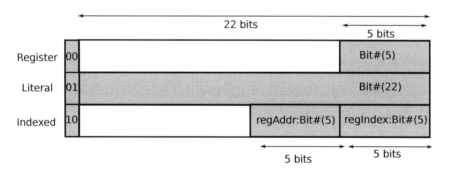

Figure 10.2: InstrOperand Tagged Union

overlays them, aligned on the LSB (to the right). Then, it tacks on as many bits at the MSB end (to the left) as is needed to represent the tags uniquely. The tag values are assigned 0, 1, 2, ... in their textual order of appearance.

Of course, instead of using `deriving(Bits)`, we could define a different representation explicitly by providing an instance declaration making `InstrOperand` an instance of the `Bits` typeclass. In this example, one could define a slightly more efficient representation, using a 1-bit tag value to distinguish Literals and non-Literals, and then using one of the unused 22 bits to distinguish Registers from Indexeds.

In the following example, a File is either an MCD *or* FD file type *or* invalid.

```
typedef union tagged {
      void      InvalidFile ;
      Bit#(31) MCD;
      Bit#(31) FD;
   } File;
```

The reason that BSV only uses tagged unions, as opposed to the "bare" unions found in C, C++ and SystemVerilog, is, once again, for type safety. A particularly insidious bug with bare unions is to create a value assuming it is one member, but to read it as if it were another member; one gets garbage bits without realizing it. By including a tag that identifies the current member type, we can provide syntactic (and type-checked) means to ensure that it is *impossible* to "misinterpret" the bits like that.

To update a tagged union variable use normal assignment notation, that is, replace the current value in the tagged union variable by an entirely new tagged union value. In a struct it makes sense to update a single member and leave the others unchanged, but in a union, one member replaces another.

Example using the `InstrOperand` tagged union:

```
InstrOperand   orand;

orand = tagged Indexed { regAddr:3, regIndex:4 };
orand = tagged Register 23;
```

A common example of a tagged union is the `Maybe` type:

```
typedef union tagged {
      void  Invalid;
      a     Valid;
   } Maybe#(type a)
      deriving (Eq, Bits);
```

In hardware, a `Maybe` is one bit longer than the member type, to hold the valid/invalid tag.

10.1.5   Maybe type

In hardware design one frequently encounters the following situation:

- A value of some type `t`

- An accompanying *valid* bit indicating whether the value is meaningful or not.

For example, this might be the input to a module, with the valid bit indicating on which cycles the environment is actually feeding a value. Or, a function/module that searches a data structure for a target may return Invalid if the target is not present, or Valid and the target value if it is present.

The BSV library provides the type Maybe#(t) for this situation. The Maybe type is an instance of a tagged union used for tagging values as either *Valid* or *Invalid*. If the value is *Valid*, the value contains a data type of data_t.

```
typedef union tagged {
      void    Invalid;
      data_t  Valid;
  } Maybe #(type data_t) deriving (Eq, Bits);
```

Two functions are defined for the Maybe type in the BSV library:

- isValid returns True if the Maybe argument is valid

- fromMaybe extracts the valid value from a Maybe argument. If the tag is invalid, a supplied default value is returned.

Examples of defining and setting a Maybe#(int):

```
Maybe#(int) m1 = tagged Valid 23;      //valid bit True, value 23
Maybe#(int) m2 = tagged Invalid;       //valid bit False, value irrelevant
```

You can use the matches statement to check the validity of a Maybe variable m, while retrieving the value. This technique can also be used in if statements and method conditions.

```
    rule onValid (m matches tagged Valid .d);
       x <= d;
```

The matches form plays two roles. First, it provides a boolean indicating whether the value matches the pattern or not, and this is used in the rule condition. Second, if there is a match, it binds the new variable d to the member value, which can be used in the rule body.

If a rule only fires when the Maybe value m is Invalid, you only have to check the tag, not the value.

```
    rule onInvalid (!isValid(m)));
       . . .
```

or

```
        rule onInvalid (m matches tagged Invalid);
           ...
```

The `isValid` function can be used to implement an if/else.

```
    if isValid (m)
        ...
    else
        dout = 0;
```

10.1.6 Case statements and pattern-matching on tagged unions

Pattern-matching provides a visual and succinct notation to compare a value against a tagged union and to access members of a tagged union. (Actually, pattern-matching is also used on structs, enums and other scalars, but here we focus on its use for tagged unions.)

A pattern-matching case statement contains the keyword `matches`. Each case item has a left-hand side and a right-hand side, separated by a colon. If the pattern-match on the left-hand side is `True`, the expression on the right-hand side of the statement is executed. On a successful match, the left-hand side may also bind new variables to the values of members of the matched value (i.e., they implicitly select those members), and these variables can be used on the right-hand side of the case item. Only the first matching statement is executed. If none of the case items succeed, the default item is selected. If there is no default item, no statement is executed.

Integer literals can include the wildcard character ?. An integer literal containing a wildcard will match any constant obtained by replacing each wildcard character by a valid digit. For example, 'h12?4 will match any constant between 'h1204 and 'h1294.

Let's look at the tagged union defined in the simple processor example in the next section, 10.2, describing the instruction set of the processor.

```
typedef union tagged {
    struct { RegNum rd; Value v; }                MovI;
    InstructionAddress                            Br;
    struct { RegNum rs; InstructionAddress dest; } Brz;
    struct { RegNum rd; RegNum rs1; RegNum rs2; } Gt;
    struct { RegNum rd; RegNum rs1; RegNum rs2; } Minus;
    RegNum                                        Output;
    void                                          Halt;
} Instruction deriving (Bits);
```

The simple processor module contains a single rule that encapsulates fetching and executing an instruction (whenever the processor is `running`). The instruction memory is modeled using an array of registers containing `Instructions` called `imem`:

```
Reg#(Instruction) imem[imemSize];
```

An excerpt of the rule is shown here:

```
rule fetchAndExecute (running);
    let instr = imem [pc];
    case (instr) matches
        tagged MovI  { rd: .rd1, v: .v1 }   : begin
                                                regs[pack(rd1)] <= v1;
                                                pc <= pc + 1;
                                              end
        tagged Br     .d                    : pc <= d;
        ...
    endcase
    ...

endrule
```

The first line in the rule body defines a name `instr` for the current instruction, `imem[pc]`.

The `case (instr) matches....endcase` phrase is the pattern-matching case statement. The statement tries to match the value in `instr` to the different members of the tagged union `Instruction`. When it finds a match, the right-hand side statement executes the statements for that instruction. For example, the first two arms of the case statement can be read in English as follows:

```
If 'instr' is of the 'MovI' kind,
    its contents are a struct with members 'rd' and 'v'
    (of type RegNum and Value, respectively); let us bind these
    members to the variables 'rd1' and 'v1' respectively,
        and then let us assign v1 to the rd1'th register
        and increment pc by 1

If 'instr' is of the 'Br' kind,
    its contents are of type 'InstructionAddress'; let us bind it
    to the variable 'd',
        and then let us assign 'd' to 'pc'
```

In each case arm, the phrase before the colon : is the *pattern* that is to be matched against `instr`. If the pattern successfully matches, it binds certain contents of `instr` to certain variables. When the pattern is successfully matched, the case arm is selected, and we execute the right-hand side (after the :) using the variables bound during the pattern matching.

As you can see, this pattern-matching notation dramatically improves readability, compared to the more traditional explicit member selection notation in C/C++ (and improved readability usually improves correctness).

Once you understand the pattern-matching, the right-hand sides of each case arm are practically self-explanatory.

10.2 Simple Processor Model

Source directory: `adv_types/simple_proc`

Source listing: Appendix A.9.2

This section describes a program for a Simple Processor computing the GCD (Greatest Common Divisor) of two numbers x and y. The example illustrates the use of enums, tagged unions, structures, arrays of registers, and pattern matching.

The program encodes Euclid's algorithm:

```
// pseudo-code
while (y != 0)
    if (x <= y)
        y = y - x
    else swap (x <=> y)
```

Excerpts illustrating specific concepts are presented in this section. The complete program is found in Appendix A.9.2. There are two files: the processor `Simple Processor.bsv` and a testbench `Tb.bsv`.

Let's look at `SimpleProcessor.bsv`. The module is very readable because the type system allow us to work with symbolic names throughout, instead of getting bogged down with the details about the bit fields in an instruction coding.

The statement:

```
typedef enum {R0, R1, R2, R3} RegNum
        deriving (Bits);
```

defines four symbolic names for the registers in a four-register register file. The `deriving (Bits)` phrase tells the compiler to pick a default bit representation (here 0, 1, 2, and 3) but we will never have to be aware of these details in the code. The statement:

```
Integer regFileSize = exp (2, valueof (SizeOf#(RegNum)));
```

calculates the size of the register file (here it is 4). The statement `Sizeof#(RegNum)` evaluates to the numeric type 2, because that is the number of bits needed to represent a `regNum`. `valueof(2)` converts this numeric type 2 into the ordinary Integer 2. Finally, `exp(2,2)` computes the value of 2 raised to the power 2, yielding the Integer 4. Thus, if tomorrow we decided to double the sizeof the register file:

```
typedef enum {R0, R1, R2, R3, R4, R5, R6, R7} RegNum
```

the system will automatically recompute `regFileSize` to be 8. The next set of statements:

```
typedef 5 InstructionAddressWidth;
typedef UInt#(InstructionAddressWidth) InstructionAddress;
Integer imemSize = exp (2, valueof(InstructionAddressWidth));
```

defines instruction addresses to be 5 bit wide, defines the `InstructionAdress` type to be a synonym for an unsigned integer of width 5, and finally computes the instruction memory size to be 2^5 yielding 32 (this is a very small computer!). Our simple processor computes with 32-bit values:

```
typedef Bit#(32)  Value;
```

The next statement defines the tagged union data structure `Instruction` (we also discussed this in the last section):

```
typedef union tagged {
    struct { RegNum rd; Value v; }                    MovI;
    InstructionAddress                                Br;
    struct { RegNum rs; InstructionAddress dest; }  Brz;
    struct { RegNum rd; RegNum rs1; RegNum rs2; }  Gt;
    struct { RegNum rd; RegNum rs1; RegNum rs2; }  Minus;
    RegNum                                            Output;
    void                                              Halt;
} Instruction deriving (Bits);
```

Notice that we do not get bogged down into details of how these members are laid out in memory, i.e., the bit representations (we will revisit this question later). The `deriving (Bits)` phrase tells the compiler to pick a default bit representation. In the module `mkSimpleProcessor`, we represent instruction memory using an array of registers, where the array size is `imemSize`. The for-loop initializes each element of the array with a fresh register using the `mkRegU` module, an uninitialized register.

```
Reg#(Instruction) imem[imemSize];
for (Integer j = 0; j < imemSize; j = j + 1)
    imem [j] <- mkRegU;
```

We looked at an excerpt of the rule `fetchAndExecute` in the previous section in our discussion of pattern-matching. Let's look at another excerpt containing all the matching statements, but not all the details for the statement execution:

```
let instr = imem [pc];
case (instr) matches
   tagged MovI  { rd: .rd, v: .v }               : begin
                                                      regs[pack(rd)] <= v;
                                                      pc <= pc + 1;
                                                  end
   tagged Br    .d                               : pc <= d;
   tagged Brz   { rs: .rs, dest: .d }            : if (regs[pack(rs)] == 0)
                                                      pc <= d;
                                                  else
                                                      pc <= pc + 1;
   tagged Gt    { rd:.rd, rs1:.rs1, rs2:.rs2 } : begin ... end

   tagged Minus { rd:.rd, rs1:.rs1, rs2:.rs2 } : begin ... end

   tagged Output .rs                             : begin ... end

   tagged Halt                                   : begin ... end

   default: begin
        display ("%0d: Illegal instruction at pc %0d: %0h", cycle, pc, instr);
        running <= False;
        end
endcase
cycle <= cycle + 1;
```

Each pattern matching phrase contains the rules for the different instructions. If none of the matches are `True`, the default clause reports an illegal instruction. Why do we use `pack` in `regs[pack(...)]` expressions? The reason is that variables like `rd`, `rs1`, `rs2`, etc. are of type `RegNum` whereas arrays are indexed by numeric types such as integers, bits, and so on. By applying `pack`, we convert a `RegNum` type into a `Bits#()` type, which is then legal as an array index.

10.3 Structs of registers vs. register containing a struct

Source directory: `adv_types/struct_reg`

Source listing: Appendix A.9.3

In this example we're going to look at the difference between a struct containing registers and a register containing a struct. Let's define two structures. The first is a struct of bit-able types, meaning we can add the provisos `deriving(Bits, Eq)` and the structure will have a default pack/unpack defined for it. This allows us to load values of this type into FIFOs, registers, etc. The `Eq` part will define `==`, so that two structures of the same type can be compared (the default is a complete bit-for-bit comparison).

```
typedef struct {
    int   a;
    Bool  b;
    } TEx1 deriving(Bits,Eq);
```

Our second structure is a struct of registers. We can't add `deriving(Bits, Eq)`, because an interface type is a static concept, and we're not trying dynamically to represent interfaces as bits, nor compare them for equality.

```
typedef struct {
    Reg#(int)  ra;
    Reg#(Bool) rb;
    } TEx2;
```

We can define a register that contains values of type TEx1, meaning that `r1` is a register value and can be read or written like any other register. Its members are accessed via its interface, but the whole entity is a register. Remember, if any bit of a register is updated, the entire register is considered updated.

```
Reg#(TEx1)    r1 <- mkReg( unpack(0) );
```

Since the structure TEx1 is just a type, we can use it anywhere we use a type that is in the Bits class (has pack/unpack defined), for example, in a FIFO.

```
FIFO#(TEx1) fifo <- mkFIFO;
```

TEx2 has interfaces defined inside of it. So we have to be careful to attach instances to the interfaces, or assign them appropriately to other interfaces which are connected to instances.

```
TEx2 r2;
r2.ra <- mkReg(2);
r2.rb <- mkReg(True);
```

We can also define registers and assign the interfaces to the same instances as defined by this interface/instance.

```
Reg#(int)  reg3A <- mkReg(3);
Reg#(Bool) reg3B <- mkReg(False);

TEx2 r3;
r3.ra = reg3A;
r3.rb = reg3B;
```

The last three lines could also have been written:

```
TEx2 r3 = TEx2 { ra: reg3A, rb: reg3B };
```

or:

```
let r3 = TEx2 { ra: reg3A, rb: reg3B };
```

A register is always updated in a single Action that updates the whole value, even if that value happens to be a struct and we logically wish to update only one of its members. To update only one of its members, you have to read out the entire struct value, compute a new struct value where the member is changed, and write back the entire new struct value.

```
rule init ( step == 0 );
    TEx1 t = r1;    // read out the entire struct value
    t.a = 20;       // compute a new struct value with changed member
    r1 <= t;        // write back the entire new struct value

    step <= step + 1;
endrule
```

In a structure of registers, on the other hand, each register can be updated individually because they have different interfaces. So you don't have to create a temp local variable.

```
rule step1 ( step == 1 );
    r2.ra <= 31;
    r2.rb <= False;

    step <= step + 1;
endrule
```

Another way to think about it is this: TEx1 represents dynamic values, i.e., a struct that exists at run time on wires, in registers, on interfaces, and so on. On the other hand, TEx2 is just a static struct, a convenient notation to describe this pair of registers as a logical group.

10.4 Tuples

Source directory: `adv_types/tuple`

Source listing: Appendix A.9.4

Tuples are predefined structures which group a small number of values together. BSV pre-defines tuples for sizes 2,3,4,5,6,7, and 8; meaning they can have 2 to 8 members. Tuples enjoy certain special syntactic support, and so You cannot define your own tuples.

The following type definition statement creates a Tuple2 type named `foo`, where the first value is a Bool and the second value is an int, and assigns values.

```
Tuple2#( Bool, int ) foo = tuple2( True, 25 );
```

As a rule of thumb, structs are usually a better choice than Tuples as each member is named within a struct. But on some occasions, you just want to collect a few values and pass them around without having to define a new struct type for this.

A set of selector functions allow you to select a specified member from the tuple. For example, `tpl_3(x)` extracts the third field from the tuple named `x`. Obviously, this must be from a tuple that has at least 3 values.

In our example, we can use `tpl_1` to read the first value of `foo`, and `tpl_2` to read the second value:

```
Bool v1 = tpl_1( foo );
int  v2 = tpl_2( foo );
```

Functions like `tuple2` are ordinary functions, and like all functions, their arguments are positional, so you cannot switch them. For example, if you tried:

```
Tuple2#( Bool, int ) foo2 = tuple2( 25, True );
```

you'd get a type-checking error, because it expects a `Bool` value in the first position and an `int` argument in the second position. This is in contrast to struct expressions where, since members are specified by name, any order is acceptable.

An example of a Tuple4:

```
Tuple4#( Bool, UInt#(4), Int#(8), Bit#(12) ) bar = tuple4( True, 0, -2, 5 );
```

Tuple components are often selected using pattern matching. In the example below, the match statement matches the left-side pattern against the values in the right-hand side expression. The variables `in` and `start` are assigned to the components of the Tuple2, defining `in` as a `Bit#(32)` and `start` as a Bool.

```
    Reg#(Bit#(32)) a <- mkReg(0);
    Tuple2#(Bit#(32), Bool) data;

    rule r1;
      match {.in, .start} = data;
      a <= in;
    endrule
```

This notation is special for tuples, and thus is not extensible by the user.

Chapter 11

Static elaboration - For-loops/while-loops

In BSV you can use `for` loops and `while` loops during static elaboration. It is important to note that this use of loops does not express time-based behavior. Instead, they are used purely as a notational means to express some repetitive code. They are statically unrolled and express the concatenation of multiple instances of the loop body statements. We first saw this behavior in Section 4.2. In particular, the loop termination condition must terminate the loop by static evaluation. Specifically, the loop condition can never depend on a value in a register or a value returned in a method call, which are only known during execution and not during static elaboration.

[BSV also has another, distint use of for- and while-loops, to express time-based (temporal) behavior of FSMs (finite state machines). These are described in Section 14.]

11.1 Non-terminating for-loops and while-loops

Source directory: `static_elab/loops`

Source listing: Appendix A.10.1

When using either type of loop, the loop iteration expressions, including the termination condition, must become known during static elaboration; it cannot depend on a dynamic value in a state element. In this example we examine what happens when you try to use a dynamic value, a value stored in a register, to determine the loop termination.

```
package Tb;

import FIFO::*;
import GetPut::*;
import ClientServer::*;

(* synthesize *)
module mkTb (Empty);

   Reg#(int) klimit <- mkReg (24);

   rule r;
      int a = 10;

      for (int k = 20; k < klimit; k = k + 1)
         a = a + k;

      $display ("rule r, a = %0d", a);
   endrule

endmodule: mkTb
endpackage: Tb
```

Let's consider the following statement in rule r:

```
for (int k = 20; k < klimit; k = k + 1)
    a = a + k;
```

In our discussion in Section 4.2, we said that loops like these are *statically unrolled,* essentially making separate copies of the loop body. The problem here is that the loop termination depends on a dynamic value, the value in register klimit. And although in this trivial example there is no other assignment to register klimit and we may infer that its value is constant at 24, in general a register's value (or any other dynamic value) is *not* constant, and therefore the compiler cannot rely on that value.

Conceptually, the compiler unrolls the loop one iteration at a time:

```
k = 20;
if (k < klimit) begin
    a = a + k;
    k = k + 1;
    if (k < klimit) begin
        a = a + k;
        k = k + 1;
```

```
            if (k < klimit) begin
                a = a + k;
                k = k + 1;
                ... and so on ...
            end
        end
    end
```

At every state, it tries to *reduce* expressions whose values are known. For example, instead of `klimit`, if we had the constant 24, the compiler would be able to reduce the first predicate:

```
(k < 24)  -> (20 < 24) ->  True
```

and then use this to eliminate the outermost `if-then-else` and so on. In particular, by repeated application of such reduction, the compiler can unroll the loop into the appropriate number of loop bodies.

However, when the predicate depends on a dynamic value (i.e., a value that is not known until execution time, because, for example, it is read out of a register), the compiler cannot reduce these expressions statically. It thus goes on forever, trying to unwind the loop until termination, which, of course, is never reached. When we compile, we see a warning message like this:

```
Warning: "Prelude.bs", line 2469, column 0: (G0024)
    The function unfolding steps interval has been exceeded when unfolding
    'Prelude.primFix'. The current number of steps is 100000. Next warning at
    200000 steps. Elaboration terminates at 1000000 steps.
```

This is a red flag that indicates that the compiler is trying to unwind a statement that may not terminate. Whenever we see a message like this, we should check if the compiler is trying to statically elaborate a loop that depends on a dynamic value.

`while` loops have the same static elaboration requirements. If you replace the `for` loop with a `while` loop like so:

```
    int k = 20;
    while (k < klimit) begin
        a = a + k;
        k = k + 1;
    end
```

you will see the same compiler behavior.

11.2   Static and dynamic conditionals

Source directory: `static_elab/conditional`

Source listing: Appendix A.10.2

BSV does not distinguish notationally between statically elaborated conditionals and dynamic conditionals (muxes). The compiler tries to reduce all expressions statically as far as possible. If the predicate reduces to a constant `True` or `False`, the compiler can statically eliminate the conditional; otherwise it remains a mux that dynamically selects the value.

BSV also does not distinguish between conditional expressions (`e1?e2:e3`), if-then-else expressions, if-then-else statements, case expressions, case statements, etc. They are all treated the same by the compiler.

In the our example, the top-level statements:

```
    Bool configVar = True;
    int  x0        = 23;
```

define some top-level constants. Note, that in general, this is a more modern, type-safe and preferable way to define constants than the traditional use of macro preprocessor `define statements. The statement:

```
        int initval = (configVar ? (x0 + 10) : (x0 + 20));
```

uses a conditional expression on the right-hand side. In an expression $e_1?e_2:e_3$, e_1 must be a boolean expression. If e_1 evaluates to `True`, then the value of e_2 is returned; otherwise the value of e_3 is returned. More generally, e_1 can also include pattern matching.

Our initial statement translates to:

```
    If configVar is True
        initval = x0 + 10
    else
        initval = x0 + 20
```

Since all the variables in the conditional statement are known, the expression is statically reduced by the compiler and the statement effectively reduces to:

```
        int initval = 33;
```

Similarly, the compiler can statically reduce the conditional expression in:

```
            Reg#(int) r0 <- (configVar ? mkReg (initval) : mkRegU);
```

to be equivalent to:

```
            Reg#(int) r0 <- mkReg (33);
```

Note that we used a statically elaborated conditional to choose between alternate module hierarchies: mkReg(0) and mkRegU. If configVar is True we will instantiate mkReg, otherwise we instantiate the module mkRegU. The same idea could be written using a conditional statement (if...else) instead of a conditional expression to initialize the register to one of two modules, either mkReg(initval) or mkRegU.

```
        Reg#(int) r1;
        if (configVar)
            r1 <- mkReg(initval);
        else
            r1 <- mkRegU;
```

The above statements are statically elaborated conditionals; the condition is known during static elaboration. You can also have conditional expressions which depend on a dynamic value, as shown in this next section from the example:

```
        Reg#(int) cycle <- mkReg (0);
        ...
        rule r;
            r0 <= ((pack (cycle)[0]==1) ? (r0 + 1) : (r0 + 5));

            if (pack (cycle)[0]==1)
                r1 <= r1 + 1;
            else
                r1 <= r1 + 5;

            if (cycle > 7) $finish (0);
            cycle <= cycle + 1;
        endrule
```

In the following statement, the predicate of the conditional depends on a dynamic value, the contents of the register cycle:

```
            r0 <= ((pack (cycle)[0]==1) ? (r0 + 1) : (r0 + 5));
```

Because we need to know the dynamic value of `cycle`, the conditional expression cannot be reduced by the compiler. Instead, it becomes hardware that makes the selection dynamically. Specifically, it becomes a mux (multiplexer) that selects one of the two values, r0+1 or r0+5. In fact, if we look at the generated Verilog file (`mkTb.v`), we see that this source statement has become the following Verilog:

```
// register r0
assign r0$D_IN = cycle[0] ? r0 + 32'd1 : r0 + 32'd5 ;
assign r0$EN = 1'd1 ;
```

The statements:

```
        if (pack (cycle)[0]==1)
            r1 <= r1 + 1;
        else
            r1 <= r1 + 5;
```

are an alternative way of saying the same thing (here with `r1` instead of `r0`). Instead of using a conditional expression, we used a conditional statement. But, in fact, the generated Verilog for `r1` is the same as that generated for `r0`:

```
// register r1
assign r1$D_IN = cycle[0] ? r1 + 32'd1 : r1 + 32'd5 ;
assign r1$EN = 1'd1 ;
```

Chapter 12

Expressions

So far we've examined many of the fundamental statements in BSV including module instantiations, variable declarations, register reads and writes, and loop statements. As we've seen, many of these statements have included expressions. An expression represents a value, usually in the form of a computation or comparison, which returns a result. Since all values in BSV have types, every expression has a type.

Expressions occur on the right-hand sides of variable assignments, on the left-hand and right-hand side of register assignments, as actual parameters and arguments in module instantiation, function calls, method calls, array indexing, and so on.

There are many kinds of primary expressions, such as identifiers, literals, and system calls. Complex expressions are built using conditional expressions along with unary and binary operators. The supported unary and binary operators, in order of precedence, are shown in the following table.

Unary and Binary Operators in order of Precedence		
Operator	Associativity	Comments
+ - ! ~	n/a	Unary: plus, minus, logical not, bitwise invert
&	n/a	Unary: and bit reduction
~&	n/a	Unary: nand bit reduction
|	n/a	Unary: or bit reduction
~|	n/a	Unary: nor bit reduction
^	n/a	Unary: xor bit reduction
^~ ~^	n/a	Unary: xnor bit reduction
* / %	Left	multiplication, division, modulus
+ -	Left	addition, subtraction
<< >>	Left	left and right shift
<= >= < >	Left	comparison ops
== !=	Left	equality, inequality
&	Left	bitwise and
^	Left	bitwise xor
^~ ~^	Left	bitwise equivalence (xnor)
|	Left	bitwise or
&&	Left	logical and
||	Left	logical or

Left associativity of an infix binary operator just means that an unparenthesized expression like this:

$$e1 + e2 + e3$$

is equivalent to the following parenthesized expression:

$$(e1 + e2) + e3$$

i.e., the operator "associates to the left".

12.1 "Don't care" expressions, 2-valued and 4-valued logic

Source directory: `expressions/dontcare`

Source listing: Appendix A.11.1

Verilog defines "4-valued logic", where a bit can take on any of the following four values:

- 0   (gnd, low, false, ...)

- 1   (vcc, high, true, ...)

- X   an uninitialized bit in a register, or a wire driven to both 0 and to 1.

- Z   a wire that is not driven at all, floating, tristate

The first two are the classical, pure boolean digital abstraction. The latter two are an artifact of Verilog simulation semantics (even synthesizable Verilog is usually restricted to 0s and 1s).

BSV semantics is based on the abstraction of classical, pure boolean logic, just 0s and 1s ("2-valued logic"). However, when BSV is compiled to Verilog and run on a Verilog simulator, you might see Xs due to uninitialized values.

In BSV you can also specify "don't care" values using the expression "?" (a question mark by itself). Note, this does not mean X or Z, it just means "unspecified whether it is 0 or 1", allowing the compiler to pick 0 or 1, as convenient. This accomplishes two goals: first, it documents your intent that you don't care whether it's a 0 or a 1, and second, the compiler can often exploit this freedom by synthesizing more efficient logic. For example, if one input to a multiplexer is "?", then the compiler can choose it to be the same as one of the other inputs, and therefore simplify the multiplexer.

Let's look at a quick example of how using ? can generate better logic. Consider the following logic table:

a	b	out
0	0	1
0	1	?
1	0	?
1	1	0

If we presume that ? == 0, as we might simply do, then we get the following logic:

```
out = ~(a & b)        one NAND gate
```

But, if we are more careful, this could optimize to:

```
out = ~a              one INV gate
```

It doesn't make a big impact on size for a single gate, but do this many times and the gate and timing savings can be significant.

"Don't care" values can be used by the BSV compiler itself for such optimization, or it may just pass it on to Verilog for later optimization by the RTL-to-gates synthesis tool. There are two compiler switches which determine how this is handled:

- -opt-undetermined-vals: tells the compiler to try to optimize don't cares.

- -unspecified-to n: tells the compiler to write unspecified values to n, where n is 0,1, or X. Note that Bluesim will only allow n to be 0 or 1, since it is a two-state simulator.

In a Verilog simulator, the ? cases will come out as X, which we expect the RTL-to-gates synthesizer to optimize to the proper value.

12.2 Case expressions

Source directory: `expressions/case`

Source listing: Appendix A.11.2

We examined case statements and pattern matching in Section 10.1.6. BSV also allows the case statement to be used as an expression, as demonstrated in this example.

```
Reg#(int) a <- mkReg(0);

rule init ( step == 0 );
   int val = case( a )
                  0: return 25;
                  1: return 23;
                  2: return 17;
                  default: return 64;
              endcase;
   step <= step + 1;
endrule
```

The lines from `case()` to `endcase` are now an *expression*, forming the right-hand side of the assignment statement. The semicolon at the end terminates the assignment statement.

Another interesting example uses the case expression to select between two modules to instantiate, based on a parameter or constant.

```
Bool initReg = True;

Reg#(int) rr <- case ( initReg )
                   False: mkRegU;
                   True:  mkReg(0);
                endcase;
```

The first example is a dynamic use of a case expression (it becomes a multiplexer); the second example is a static use of a case expression (is processed, resolved and disappears during static elaboration).

12.3 Function calls

Source directory: `expressions/function_calls`

Source listing: Appendix A.11.3

A function definition is introduced by the keyword `function`. Every function must define the type of the value returned by the function and a name for the function. The arguments and result of a function can be of any BSV type at all (including other function types). Let's look at two simple functions definitions:

```
function int square( int a );
   return a * a;
endfunction

function int increment( int a );
   // saturate at maxInt
   int maxInt = unpack({ 1'b1, 0 });    // a local constant
   if (a != maxInt)
      return a + 1;
   else
      return a;
endfunction
```

These function expressions can be used in statements and they can be nested since they don't (indeed cannot) update any state.

```
Reg#(int) step <- mkReg(0);

rule init ( step == 0 );
   $display("=== step 0 ===");
   $display("square    10 = ", square(10));
   $display("increment 20 = ", increment(20));

   // nested functions
   $display("2*10+1       = ", increment(square(10)));

   step <= step + 1;
endrule
```

In the above example, the functions represent combinational circuits. However, the same functions can also be used during static elaboration:

```
Reg #(int) foo = mkReg (increment (square (10)));
```

Here, the functions are statically elaborated to the value 101 which is used as the initialization value of the register. In BSV, we make no syntax distinction between the language used for static elaboration and the language used to express dynamic computations—it is always clear from context how a function is being used.

Chapter 13

Vectors

BSV provides three types to describe collections of elements, each selected by one or more indices: vectors, arrays, and lists. The Vector type, provided in the package of the same name, is the most useful. Vectors are completely synthesizable and the package contains a large number of functions to operate on the Vector data type.

We can define any type, user-defined or BSV-provided, as an array simply by using the [ ] notation. The value within the [ ] must be a constant expression, that is computable during static elaboration. An array can be multidimensional.

A Vector is an abstract data type which is a container of a specific length, holding elements of one type. To use vectors you must import the Vector package which includes many functions to create and operate on vectors. The length of a vector must be resolvable during type-checking (although it is possible to write functions that are polymorphic in the vector length). If the data type of the elements is in the Bits typeclass, the vector is in the Bits typeclass. In other words, a vector can be turned into bits if the individual elements can be turned into bits. Therefore, vectors are fully synthesizable. A vector can also store abstract types, such as a vector of rules or a vector of Reg interfaces. These are useful during static elaboration although they have no hardware implementation.

Lists are similar to vectors, but are used when the number of items in the list may vary at compile-time or need not be strictly enforced by the type system. All elements of a list must be of the same type. The List package defines a data type and functions which create and operate on lists. Lists are most often used during static elaboration (compile-time) to manipulate collections of objects when the size of the list is unknown. Lists are not synthesizable since their size may not be known.

Neither arrays nor lists are in the Bits typeclass, and therefore cannot be stored in registers or other dynamic elements. When going through a synthesis boundary, where wires must be created, arrays and lists must be converted into vectors in order to generate hardware.

The following examples define collections of elements of type UInt#(16):

```
UInt#(16) arr[10];         // an array named arr of 10 values
Vector#(10, UInt#(16) vec; // a vector named vec of 10 values
List#(UInt#(16)) lis;      // a list named lis of unknown number of values
```

13.1   Arrays and square-bracket notation

Source directory: `vector/arrays`

Source listing: Appendix A.12.1

We can define any type, user-defined or BSV-provided, as an array simply by using the [ ] notation. The value within the [ ] must be a constant expression, computable during static elaboration. Arrays can be multidimensional.

In this example, we're creating an array named `arr` with 10 instances, each of type `UInt#(16)`. Each instance of `arr` can be accessed as `arr[0]`, `arr[1]`, ..., `arr[9]`.

```
UInt#(16) arr[10];
```

Now we can initialize it statically at the global level:

```
arr[0] = 0
arr[1] = 3
arr[2] = 6
...
```

A statically-elaborated for-loop can be used to initialize an array. The following is identical to typing the 10 lines out separately:

```
for (Integer i=0; i<10; i=i+1)
    arr[i] = fromInteger(i * 3);
```

You can create multi-dimensional arrays. There is no theoretical limit on the number of dimensions, but you have to remember that you are creating logic that will be implemented in hardware for selection and update. An example of a 2-dimensional array:

```
UInt#(16) arr2[3][4];

for (Integer i=0; i<3; i=i+1)
    for (Integer j=0; j<4; j=j+1)
        arr2[i][j] = fromInteger((i * 4) + j);
```

You can create arrays of register interfaces, or any module interface, in a similar way. Remember to initialize the array, for example using a for-loop, or else it will contain a "don't care" interface of the element type:

```
Reg#(int) arr3[4];
for (Integer i=0; i<4; i=i+1)
    arr3[i] <- mkRegU;              // initialization
```

You can write to all or any of the array's registers within the same rule.

```
rule load_arr3;
    arr3[0] <= 'h10;
    arr3[1] <= 4;
    arr3[2] <= 1;
    arr3[3] <= 0;
endrule
```

13.2 Arrays vs. Vectors

Source directory: `vector/arr_vs_vector`

Source listing: Appendix A.12.2

Vectors are a much more powerful implementation of array functionality. In this section we discuss the basics of vectors, which are defined in the `Vector` package. To use vectors, you must import the package. This example also uses a FIFO, so we'll import that package as well.

```
import Vector::*;
import FIFO::*;
```

Like arrays, vectors are used to group items together, but vectors have additional features and a different syntax. The `Vector` package defines an abstract data type, a vector, which is a container of a specific length, holding elements of one type. Functions which create and operate on vectors are also defined within the `Vector` package.

The vector type is written: `Vector#(`*number of items, type of items*`)`

The function `newVector` can be used to create a new vector value with unspecified contents:

```
Vector#(4, Int#(16)) vec1 = newVector;
```

The Vector package provides many functions for creating and initializing a vector, including the following functions:

- `newVector` creates a vector value with n unspecified elements; typically used when a vector is declared.

- genVector creates a vector value with index 0 containing Integer 0, index 1 containing Integer 1, ..., index $n - 1$ containing Integer $n - 1$.

- replicate creates a vector of n elements by replicating the given argument c at every index

- genWith creates a vector of n elements by applying the given function to Integer arguments 0 through $n - 1$. The argument of the function is another function which has one argument of type Integer and returns an element value of the vector element type. Thus, genVector can be regarded as an application of genWith to the "identity" function.

This statement uses the replicate function to create vec2 and assign the value 0 to all the elements. replicate creates as many items as needed of the given argument, in this case, 4 items of type Int#(16).

```
Vector#(4, Int#(16)) vec2 = replicate( 0 );
```

This is essentially the same functionality as using arrays:

```
Int#(16) arr2a[4];
for (Integer i=0; i<4; i=i+1)
    arr2a[i] = 0;
```

You can read and write vec2 and arr2a the same way, but the definition of the vector notation of vec2 takes only one line.

We can also use the map function, which allows you to apply a function to all elements of a vector. The map function takes as an argument a function, which is then applied to each element of the vector given as its second argument, and returns a new vector. For example:

```
Vector#(4, Int#(16)) vec3 = map( fromInteger, genVector );
```

Here, genVector creates a vector of 4 elements (0,1,2,3), each of type Integer. Then, map applies the function fromInteger to each element, returning a vector of four elements (0,1,2,3), each of type Int#(16). This becomes the initial value of the declared variable vec3. Conceptually, it is equivalent to the following "unrolled" form:

```
vec3[0] = fromInteger(0)
vec3[1] = fromInteger(1)
vec3[2] = fromInteger(2)
vec3[3] = fromInteger(3)
```

With arrays, you'd have to do it with an explicit loop:

```
Int#(16) arr3a[4];
for (Integer i=0; i<4; i=i+1)
    arr3a[i] = fromInteger(i);
```

We can even write our own function to load whatever we want at each index. For example, let's create a `times13` table vector. The function `times13`:

```
function Int#(16) times13 ( Integer i );
    return fromInteger(i * 13);
endfunction
```

can be applied across the vector created by `genVector`:

```
Vector#(4, Int#(16)) vec4 = map( times13,  genVector );
```

to produce a vector of 4 elements (0, 13, 26, 39), with each element of type `Int#(16)`.

This is how it would be done with arrays:

```
Int#(16) arr4a[4];
for (Integer i=0; i<4; i=i+1)
    arr4a[i] = fromInteger(i * 13);
```

Vector functions are more composable; they can be nested as deep as you wish. They encourage a more "functional programming" style. As you work with vectors and vector functions, you will see vectors have many advantages over arrays.

13.3   bit-vectors and square bracket notation

Source directory: `vector/bit_vector`

Source listing: Appendix A.12.3

Like Verilog, we can access the bits of `Bit` data types, but not of `Int` and `UInt` (at least not without recasting them to `Bit` via the `pack` function). The idea is to use `Int` and `UInt` for mathematical operations where their bit representations are an irrelevant implementation detail. For example, the mathematical operations on `Int` and `UInt` can be defined independently of whether they are represented in "little-endian" or "big-endian" forms of bits. That being said, sometimes there are hardware-related data-specific tricks with numbers, like *is this number divisible by 4* or `bits[1:0]==0`. Regardless, you should try to use `Bit` only for things that are bit type structures and not for counters, etc. The more "representation-independent" your code is, the more it will be robust to changes in representation.

Accessing bits If you define a Bit type, you can use [ ] to access a particular bit.

```
rule test1;
   Bit#(5) foo1 = 5'b01001;
   $display("bit 0 of foo1 is ", foo1[0]);
   $display("bit 1 of foo1 is ", foo1[1]);
   $display("bit 2 of foo1 is ", foo1[2]);
   $display("bit 3 of foo1 is ", foo1[3]);
   $display("bit 4 of foo1 is ", foo1[4]);
   ...
```

resulting in the output:

```
                    bit 0 of foo1 is 1
                    bit 1 of foo1 is 0
                    bit 2 of foo1 is 0
                    bit 3 of foo1 is 1
                    bit 4 of foo1 is 0
```

Retrieving a a range of bits You can also get a range of bits:

```
   Bit#(6) foo2 = 'b0110011;
   $display("bits 2:0 of foo2 are %b", foo2[2:0]);
```

resulting in the output:

```
              bits 2:0 of foo2 are 011
```

Defining a bit vector of a new size This returns bit vector of a new size:

```
   Bit#(3) foo3 = foo2[5:3];
   $display("bits 5:3 of foo2 are %b", foo3);
```

resulting in the output:

```
              bits 5:3 of foo2 are 110
```

Setting bits

```
    foo3 = 3'b111;
    $display("set foo3[2:0] to 1s => %b", foo3);
    foo3[1:0] = 0;
    $display("set foo3[1:0] to 0s => %b", foo3);
```

resulting in the output:

```
            set foo3[2:0] to 1s => 111
            set foo3[1:0] to 0s => 100
```

Using a vector of bit instead of a bit vector A better way to use and think of these functions is to create a vector of bit (where bit is an alias for Bit#(1)).

```
    Vector#(8, bit) bar1 = replicate(0);
```

If you define a vector you then have access to the large set of vector processing functions defined in the Vector package. For example, let's look at the genWith function. This function generates a vector of elements by applying a given function to all the elements of a vectors. We're going to define the function isInteresting, which takes an Integer as an argument. It sets each bit to 1 if it is divisible by 3.

```
    function bit isInteresting( Integer i );
        return (i % 3 == 0) ? 1 : 0;
    endfunction
```

We can then use the genWith vector function to apply isInteresting to the vector bar2,

```
    Vector#(16,bit) bar2 = genWith( isInteresting );
    $display("bar2 is %b", bar2)
```

resulting in the output:

```
            bar2 is 1001001001001001
```

Rotating and shifting The Vector package provides functions for rotating and shifting vectors. Because 16'b0101_0001_1000_1001 is a sized literal, we need to use the unpack function to turn it into a Vector#(16, Bit#(1)) value. If Vector#(16, Bit#(1)) had an instance defined in the SizedLiteral typeclass, then the function fromSizedLiteral would be defined for it, and the unpack wouldn't be necessary.

```
Vector#(16,bit) bar3 = unpack(16'b0101_0001_1000_1001);
$display("bar3            is %b", bar3)
```

The reverse function reverses the order of the elements.

```
$display("reverse bar3 is %b\n", reverse( bar3 ));
```

The rotate function moves each element down by 1. The element at index n moves to index n-1, the zeroth element moves to index n.

```
$display("bar3            is %b", bar3);
$display("rotate1 bar3 is %b", rotate( bar3 ));
$display("rotate2 bar3 is %b", rotate(rotate( bar3 )));
```

The above displays return the following results:

```
bar3            is 0101000110001001
reverse bar3 is 1001000110001010

bar3            is 0101000110001001
rotate1 bar3 is 1010100011000100
rotate2 bar3 is 0101010001100010
```

You can also define your own functions, using the functions in the Vector package. Here we wrap rotateBy function to create rotateUp and rotateDown, where up moves each to the next higher index (0->1->2->3...n->0) and down moves each bit down an index (n->n-1->n-2...1->0,0->n).

```
function Vector#(n,td) rotateUp( Vector#(n,td) inData, UInt#(m) inx)
   provisos( Log#(n,m) );
   return rotateBy( inData, inx );
endfunction

function Vector#(n,td) rotateDown( Vector#(n,td) inData, UInt#(m) inx)
   provisos( Log#(n,m) );
   UInt#(m) maxInx = fromInteger(valueof(n)-1);
   return rotateBy( inData, maxInx-inx );
endfunction
```

The rotation amount can up to n places, which can be represented in log (to base 2) bits. The provisos assert this relationship between the vector length and the width of the rotation amount. With this functions you can rotate the bits in the vector by any specified number of places.

```
$display("bar3              is %b",    bar3);
$display("rotate up 1 bar3 is %b",    rotateUp( bar3, 1 ));
$display("rotate up 2 bar3 is %b",    rotateUp( bar3, 2 ));
$display("rotate up 3 bar3 is %b\n", rotateUp( bar3, 3 ));

$display("bar3              is %b",    bar3);
$display("rotate dn 1 bar3 is %b",    rotateDown( bar3, 1 ));
$display("rotate dn 2 bar3 is %b",    rotateDown( bar3, 2 ));
$display("rotate dn 3 bar3 is %b",    rotateDown( bar3, 3 ));
```

This results in the output:

```
              bar3              is 0101000110001001
              rotate up 1 bar3 is 1010001100010010
              rotate up 2 bar3 is 0100011000100101
              rotate up 3 bar3 is 1000110001001010

              bar3              is 0101000110001001
              rotate dn 1 bar3 is 0101010001100010
              rotate dn 2 bar3 is 0010101000110001
              rotate dn 3 bar3 is 1001010100011000
```

Note that in this section we are describing bit vectors, where the elements of the vector are all `Bit#(1)`. The vector functions are defined for any element type. For example, the `rotate` function moves elements of any data type. The `rotateBitsBy` function is only defined for bit vectors.

You can still quickly pack and unpack bit vectors to a `Bit#()`.

```
Bit#(16) val = 25;
bar3 = unpack( val );
$display("bar unpacked = %b / %d", bar3, bar3);

bar3[7] = 1;

// and back
Bit#(16) val2 = pack( bar3 );
$display("bar packed   = %b / %d", val2, val2);
```

The output is:

```
              bar unpacked = 0000000000011001 /    25
              bar packed   = 0000000010011001 /    153
```

13.4   Whole register update

Source directory: vector/whole_register

Source listing: Appendix A.12.4

In BSV, registers are considered atomic units with regard to their updates. That is, if any bit of a register is updated, then the entire register is considered updated.

Let's define registers containing bit vectors:

```
Reg#(Bit#(8)) foo  <- mkReg(0);
Reg#(Bit#(8)) foo2 <- mkReg(0);
```

Reading any number of bits in any of the various bit-selection forms is fine:

```
rule up1;
    $display("foo[1:0] = %b", foo[1:0]);
    $display("foo[2]   = %b", foo[2]);
```

Similarly, writing single bits within a rule is fine.

```
rule up2;
    foo[1] <= 1;
```

However, BSV considers this register to be a single entity, and this is just an abbreviation for a read-modify-write where the entire value is read out of the register, a new value is computed with the changed bit, and the entire value is written back. Thus, if we tried to write the following:

```
rule update1;
    foo[1] <= 1;
    foo[2] <= 1;
```

we will get the following error:

```
Error: "Tb.bsv", line 14, column 9: (G0004)
Rule 'RL_update1' uses methods that conflict in parallel:
  foo.write(...)
and
  foo.write(...)
For the complete expressions use the flag '-show-range-conflict'.
```

because it is regarded as two read-modify-writes of the entire register, which cannot be done as part of a single atomic Action.

If you really need to update multiple bits in a register in a single rule, couch it as a single read-modify-write: read out the whole value, construct a new value with the multiple bits changed, and write back the entire new value:

```
let tmp = foo2;     // single read into a temporary variable
tmp[1] = 1;
tmp[3] = 1;
tmp[7:6] = 3;
foo2 <= tmp;        // single write
```

This also applies to bits across rules. One rule can't update foo[1] while another rule updates foo[2]. Let's consider the following two rules:

```
rule up1;
   foo[1] <= ~foo[1];
endrule

rule up2;
   foo[2] <= ~foo[2];
endrule
```

In this case, up1 and up2 have a resource conflict on the register foo. Even though foo has 8 bits and the rules are writing different bits, BSV considers the register to be one indivisible piece of state.

If you really need to do this, consider creating separate bit registers and concat the bits together when you read or write the entire register.

```
    Reg#(bit) b1 <- mkReg(0);
    Reg#(bit) b2 <- mkReg(0);

    rule up3;
        b1 <= ~b1;
        b2 <= ~b2;
    endrule

    rule done (b2 == 1);
        $display( "b12 = %b", {b1,b2} );
        $finish;
    endrule
```

13.5   Vector of registers

You can create vectors of interfaces, including register interfaces. (Recall that in BSV, registers are just another module with two methods, _write and _read.)

You can use a one line statement to define a vector of registers using the vector function replicateM. If you accidently use replicate instead of replicateM (think of the M as replicate *Module*) you will get the following error message:

```
    Error: "Tb.bsv", line 20, column 35: (T0107)
        The type 'Vector::Vector#(12)' is not a module. Perhaps you have used the
        wrong function in a module instantiation, such as 'map' vs 'mapM'.
        Consider the modules at the following locations:
            "Tb.bsv", line 20, column 46
```

For now, just remember:

```
    ....... <- replicateM( )    // notice <-
    ....... =  replicate( )     // notice =
```

The difference is explained by the fact that replicateM must perform a side-effect (instantiation of a module) for each element, and return the results, whereas replicate does not have any side effects.

The following statement defines a vector of registers containing ints, using the replicateM function.

```
    Vector#(12, Reg#(int)) arr1 <- replicateM( mkReg(0) );
```

13.6   Register containing a vector vs. a vector of registers

Source directory: `vector/regvec_vecreg`

Source listing: Appendix A.12.5

- Use a *register containing a vector* when you will be updating the entire vector together, at once, in a single rule (atomically).

- Use a *vector of registers* when you will be updating a single element at a time.

The main issue is that when one field in a register is updated, the entire register is updated, even if most of the fields do not change.

Assignments to a register containing a large vector must maintain the values for any unchanged fields. This can lead to increased mux size, since all elements of the vector must be updated, even if only a single element has changed.

On the other hand, if you have a vector of registers, and you dynamically pick a single register to update, the scheduler may think you are touching all the registers in the vector (unless it can statically know which index you are using). This can lead to rule conflicts where none may in fact exist.

Let's look an an example to compare. In both, we're going to have an enum, named `Status` and a rule to `updateStatus`

```
import Vector :: *;
...
typedef enum {  OK,
                FAIL,
                TEST,
                WAIT
        } Status  deriving (Bits, Eq);
```

Let's define two different register/vector structures:

- `regstatus` is a register containing a vector of 10 status values (register containing a vector)

- `vecstatus` is a vector containing 10 registers of status (vector of registers)

We also define a register containing the failure condition where the *Valid* value is the value of type `Status`.

```
(* synthesize *)
module mkTb (Empty);
   Reg#( Vector#(10,Status) ) regstatus <- mkReg( replicate(OK) ) ;
   Vector#( 10, Reg#(Status) ) vecstatus <- replicateM (mkReg (OK)) ;

   Reg#(Maybe#(UInt#(4))) failCond <- mkReg(Invalid);
   Reg#(UInt#(4)) cycle <- mkReg(0);
```

To update elements in a register containing a vector (Reg#(Vector#(n, element))), you have to read out the entire value of the register (a vector), update the relevant elements, then write back the new value (a vector).

```
rule updateRegStatus (failCond matches tagged Valid .chan );
   let tempstatus = regstatus;     // read out whole vector value
   tempstatus[chan] = FAIL;
   tempstatus[chan + 1] = TEST;
   regstatus <= tempstatus;        // writeback whole vector value
endrule
```

In the vector of registers, you can have multiple statements updating the vector, since each is writing a different, single register.

```
rule updateVecStatus (failCond matches tagged Valid .chan );
   vecstatus[chan] <= FAIL ;
   vecstatus[chan + 1] <= TEST;
endrule
```

13.7   Vector of module interfaces

Source directory: `vector/modules`

Source listing: Appendix A.12.6

In the last section we discussed a vector of registers (more precisely, a vector of register interfaces). This is possible because we can express vectors of anything, including interfaces. Let's look at a small 2x4 switch, consisting of 2 vectors of FIFOs. `ififo` is a vector of 2 interfaces of type FIFO. `ofifo` is a vector of 4 interfaces of type FIFO.

```
// two fifos in
Vector#(2,FIFO#(TD)) ififo <- replicateM( mkFIFO );

// four fifos out
Vector#(4,FIFO#(TD)) ofifo <- replicateM( mkFIFO );
```

We'll take a simple approach and create a rule for each input FIFO to drive data to the output FIFO. The `preempts` attribute removes a warning by ensuring that if `move0` fires then `move1` does not.

`ififo[0]` is a single interface of type `FIFO`, which has `first`, `deq`, `enq`, and `clear` methods defined. We can access those methods directly.

```
rule move0;
   let data = ififo[0].first();
   ififo[0].deq();

   case ( data[13:12] )
      0: ofifo[0].enq( data );   // call (ofifo[0]) method enq( data )
      1: ofifo[1].enq( data );
      2: ofifo[2].enq( data );
      3: ofifo[3].enq( data );
   endcase
endrule

rule move1;
   let data = ififo[1].first();
   ififo[1].deq();
   case ( data[13:12] )
      0: ofifo[0].enq( data );
      1: ofifo[1].enq( data );
      2: ofifo[2].enq( data );
      3: ofifo[3].enq( data );
   endcase
endrule
```

If you run the complete example in appendix A.12.6, you will see how we create 4 rules to watch each output and can view the data from each output.

13.8   Static and dynamic indexing of vectors

Source directory: `vector/static_elab`

Source listing: Appendix A.12.7

Static indexing means that the compiler can determine *at compile time* the values needed to index an array. This is usually done with an index of type `Integer`, but can be done with other data types as well. For instance, let's create a vector of registers and look at various ways we can read its contents.

```
    Vector#(8,Reg#(int)) arr1 <- replicateM( mkRegU );

    Reg#(int) step <- mkReg(0);
```

First, we'll index over a fixed predetermined range (0...7).

```
    rule init ( step == 0 );
        for (Integer i=0;  i<8;  i=i+1)
            arr1[i] <= fromInteger(i);
        step <= step + 1;
    endrule
```

The indexes are all known during static elaboration, and so the compiler can easily determine what actions to take. This essentially expands statically to:

```
        arr1[0] <= fromInteger(0);
        arr1[1] <= fromInteger(1);
        arr1[2] <= fromInteger(2);
              ::          ::
        arr1[7] <= fromInteger(7);
```

This analysis holds even if the index were some type other than `Integer`, as long as the compiler can figure it out statically, such as:

```
    rule step1 ( step == 1 );
        for (int i=0;  i<8;  i=i+1)
            arr1[i] <= i;
        step <= step + 1;
    endrule
```

The variable i here is an `int`, which is shorthand for `Int#(32)`. Since this is the same data type as the elements of `arr1`, we don't need `fromInteger`, instead we can assign the value of i directly to `arr[i]`.

Also notice that there is no dependence between the steps of this loop, i.e., it is just a notation for a big parallel assignment into multiple registers.

In this next rule, we'll make it a bit more interesting. For instance, suppose we want to shift the elements of this array manually:

```
rule step2 ( step == 2 );
   for (int i=0; i<8; i=i+1)
      if (i != 7)
         arr1[i+1] <= arr1[i];
      else
         arr1[0]   <= arr1[7];
   step <= step + 1;
endrule
```

The compiler is still able to statically elaborate this loop into the desired rotation. Variable i is still statically incrementing from 0...7. Observe that now we have an if-else structure (and we could have used any case or structure that elaborates properly). This loop simply elaborates to:

```
arr[1] <= arr[0]
arr[2] <= arr[1]
arr[3] <= arr[2]
arr[4] <= arr[3]
arr[5] <= arr[4]
arr[6] <= arr[5]
arr[7] <= arr[6]
arr[0] <= arr[7]
```

To scramble the values arbitrarily, we could use a case statement, and it's still static.

```
rule step3 ( step == 3 );
   for (int i=0; i<8; i=i+1)
      case (i)
         0: arr1[0] <= arr1[3];
         1: arr1[1] <= arr1[1];
         2: arr1[2] <= arr1[2];
         3: arr1[3] <= arr1[0];
         4: arr1[4] <= arr1[5];
         5: arr1[5] <= arr1[7];
         6: arr1[6] <= arr1[4];
         7: arr1[7] <= arr1[6];
   step <= step + 1;
endrule
```

Now let's look at dynamic (run-time) indexing. Example:

```
    Reg#(UInt#(3)) inx <- mkRegU;

rule step4 ( step == 4 );
   arr1[inx] <= 0;

   step <= step + 1;
endrule

... // other rules and methods that modify inx
```

Now the array is indexed by `inx` which is the `UInt#(3)` contents of a register. This creates a runtime index, meaning when this rule fires, whatever value is in register `inx` will be used to index to the proper array value. Therefore it must create the appropriate logic to be able to write every `arr1` depending on the value of `inx`.

Now, as an optimization, if you *know* somehow that you only ever write to `inx==0` and `inx==1` (for instance), then you can give the compiler a clue by coding something like this:

```
//    if ( inx == 0 )
//       arr1[0] <= 0;
//    else if ( inx == 1 )
//       arr1[1] <= 1;
```

This gives the compiler more information, resulting in better logic.

Chapter 14

Finite State Machines (FSMs)

Finite state machines (FSMs) are common in hardware design. First, we'll examine how to build an FSM using rules. Rules can be used to build arbitrary FSMs. Then we'll demonstrate the StmtFSM sub-language which is simpler to use for "well-structured" FSMs, i.e., FSMS built involving sequencing, parallelism, conditions and loops and which have a a precise compositional model of time. The StmtFSM sub-language is simpler in these cases because it provides a succinct notation and automates all the generation and management of the actual FSM state. However StmtFSM does not introduce any new semantics—everything is explained in terms of rules, and in fact that is how the BSV compiler implements it.

14.1 Building an FSM explicitly using rules

Source directory: `fsm/rules`

Source listing: Appendix A.13.1

In this section we look at how to create a finite state machine (FSM) using rules. We define a rule for each state. The body of the rule performs the state's actions, and also updates the state to effect a state transition. If we were using standard RTL, we would have to mux all the possible values for each state to a single mux and then generate all the select signals. With BSV, the compiler generates all the muxing and control for us. This means that each rule is self-contained and much easier to understand. Need to add another state? Just add another rule. Need to add another exit arc from a state? Just add it to the if/else conditions that updates the state in the rule.

All the usual concepts of rules apply. Implicit (or ready) conditions will cause a rule not to fire. Additionally, the explicit (or rule) condition can be as complicated as you desire. The end effect is that when the rule fires, you specify the state update and actions for the rule.

We'll examine in later examples the StmtFSM sub-language, another way to implement state machines that is simpler to use in many cases. But rules are the basic, most powerful way to implement a state machine.

An enum is a good way to ensure that we have meaningful and mnemonic state names. Since it is strongly-typed, it is difficult to misuse. We declare a register to hold the current state, and initialize it to IDLE.

```
package Tb;

typedef enum { IDLE, STEP1, STEP2, STOP } State deriving(Bits,Eq);

(* synthesize *)
module mkTb (Empty);

   Reg#(State)      state <- mkReg(IDLE);
```

We declare some some additional registers that participate in the state transition decisions. The counter runs for 200 cycles.

```
   Reg#(int)      counter <- mkReg(0);
   Reg#(Bool)     restart <- mkReg(False);

   rule runCounter;
      if (counter == 200) begin
         $display("Done");
         $finish;
      end
      counter <= counter + 1;
   endrule
```

We then have one rule for each state. The following rule represents the behavior in the IDLE state. In this example, the only action is to display a message, which it does repeatedly until the counter is a multiple of 4 (4,8,12, etc), at which point it transitions to the STEP1 state.

```
   rule stateIdle ( state == IDLE );   //default state
      $display("Counter = %03d, State: IDLE", counter);
      if (counter % 4 == 0)
         state <= STEP1;
   endrule
```

This rule represents state STEP1, where it remains until it either sees the restart signal in which case it returns to the IDLE state, or it observes that the counter is a multiple of 8, when it transitions to state STEP2.

```
   rule stateStep1 ( state == STEP1 );
      $display("Counter = %03d, State: STEP1", counter);
      if (restart)
         state <= IDLE;
      else if (counter % 8 == 0)
         state <= STEP2;
   endrule
```

In state STEP2 it immediately transitions further to the STOP state.

```
   rule stateStep2 ( state == STEP2 );
      $display("Counter = %03d, State: STEP2", counter);
      // advance to the next state, ignore restart signal here
      state <= STOP;
   endrule
```

In the STOP state, it resets the state back to IDLE.

```
   rule stateSTOP ( state == STOP );
      $display("Counter = %03d, State: STOP", counter);
      state <= IDLE;
   endrule
 endmodule
 endpackage
```

In summary, coding an FSM using rules is simple: each rule condition represents a particular state. The body of a rule performs the actions necessary in the state, and also performs updates that affect rule conditions, i.e., enables the rule for the next state. In this way, arbitrary state-transition diagrams can be encoded.

14.2 One-hot FSM explicitly using rules

Source directory: fsm/onehot

Source listing: Appendix A.13.2

In this section we will repeat the previous example, except that we shall use "one-hot" encoding for the state. In the previous example, the state was expressed using an enum with four labels, encoded in two bits as 00, 01, 10 and 11. In one-hot encoding, instead, the state is encoded in four bits as 0001, 0010, 0100 and 1000, respectively. Notice that in each encoding, exactly 1 bit is set, hence the name "one-hot".

One-hot encoding is more expensive in state bits. Whereas normal state encoding (like the enum) takes $\log_2(n)$ bits to represent n state values, one-hot coding takes n bits. The payoff comes in decoding—with one-hot coding, we only need to test one bit to recognize a particular state, instead of testing $\log_2(n)$ bits. In some performance-critical situations (tightly timed hardware), this difference is crucial.

```
(* synthesize *)
module mkTb (Empty);

   // Indexes of bits for each state
   Integer idle  = 0;
   Integer step1 = 1; // or "idle + 1"
   Integer step2 = 2;
   Integer stop  = 3;

// Rather then write down the one-hot codings explicitly where it is easy to get
//   it wrong, we use a function to create them

   // create the single bit in the proper place, zero all others
   function Bit#(4) toState( Integer st );
      return (1 << st);     // left-shift '1' by st places
   endfunction

   Reg#(Bit#(4)) state <- mkReg( toState(idle) );

   Reg#(int)    counter <- mkReg(0);
   Reg#(Bool)   restart <- mkReg(False);

   rule runCounter;
      if (counter == 200) begin
         $display("Done");
         $finish;
      end
      counter <= counter + 1;
   endrule
```

The rule conditions pick out one bit to check using a constant integer index, rather than comparing against all the state bits. Otherwise, the rules (steps) are the same as in the previous example.

```
   rule stateIdle ( state[idle] == 1 );
      $display("Counter = %03d, State: IDLE", counter);
      if (counter % 4 == 0)
         state <= toState( step1 );
   endrule
```

```
    rule stateStep1 ( state[step1] == 1 );
        $display("Counter = %03d, State: STEP1", counter);
        if (restart)
            state <= toState( idle );
        else if (counter % 4 == 0)
            state <= toState( step2 );
    endrule
```

```
    rule stateStep2 ( state[step2] == 1 );
        $display("Counter = %03d, State: STEP2", counter);
        state <= toState( stop );
    endrule

    rule stateStop ( state[stop] == 1 );
        $display("Counter = %03d, State: STOP", counter);
        state <= toState( idle );
    endrule

endmodule
endpackage
```

Although we know that these rules are mutually exclusive because of the "one-hot" property of state, this mutual exclusivity is not known to the compiler, and so it will generate a number of conflict warnings and conflict resolution logic. The mutually_exclusive attribute (Section 7.2.3) informs the compiler about mutual exclusivity:

```
    (* mutually_exclusive = "stateIdle, stateStep1, stateStep2, stateStop" *)
```

14.3 StmtFSM

Another way to define FSMs is with the StmtFSM library package, an extremely powerful way of specifying structured FSMs in BSV. At the heart of this capability is a sub-language for specifying structured FSMs, built out of basic Actions. The compiler expands this sub-language into explicit rules, just as if you had written the FSM explicitly with rules with the basic Actions in the rule bodies. In other words, the sub-language is a notational convenience and does not introduce any new semantics.

The constructs in the sub-language are first-class objects, of type Stmt. You can write your own functions and static-elaboration facilities to compute with Stmt objects.

This library is extremely useful for developing testbenches, as it allows you to quickly generate C style loops, sequences, parallel blocks, etc. There is no "goto" (you can only create structured FSMs). One caveat: state encodings produced by the compiler from StmtFSM may not be as efficient as hand-coded FSMs with explicit rules that exploit semantic knowledge about the states.

14.3.1   A basic template

Source directory: `fsm/stmtfsm_basic`

Source listing: Appendix A.13.3

Let's look at a template showing the basic components of declaring, defining, and running an FSM.

1. You must import the `StmtFSM` package.

```
import StmtFSM :: *;
```

2. Create an FSM specification, which is an expression of type `Stmt`, using the sequential, parallel, conditional, and looping constructs of the `Stmt` sub-language. Entry into the scope of the FSM sub-language is signalled by the keyword `seq`.

```
Stmt test =
seq
    $display("I am now running at ", $time);
    $display("I am now running one more step at ", $time);
    $display("And now I will finish at", $time);
    $finish;
endseq;
```

3. Instantiate the FSM using the module constructor `mkFSM`, passing it the `Stmt` value above as a parameter. The state machine is automatically constructed from the procedural description given in the `Stmt` spec.

```
FSM testFSM <- mkFSM( test );
```

The module instantiation returns an FSM interface which has a **start** Action method for starting the FSM, and a **done** boolean method that can be tested for FSM completion.

4. Use the **start** method to begin state machine execution. The **start** method has an implicit condition that the FSM cannot already be running, that is, it must be **done**.

```
rule startit;
    testFSM.start;
endrule
```

5. Use the **done** boolean method to test for FSM completion.

```
rule sequel (testFSM.done);
    ...
endrule
```

Statements can be composed into sequential, parallel, conditional, and loop forms. In the sequential form (`seq...endseq`), the contained statements are executed one after the other. A statement can be an Action, in which case it forms a rule, or it can be a nested FSM. A `seq` block terminates when its last contained statement terminates; the total time is equal to the sum of the individual statement times (the number of clock cycles will be greater than or equal to the number of contained Actions, since the Actions become rules, and rules may stall according to rule scheduling semantics).

In the parallel form (par...endpar), the contained statements (threads) are all executed in parallel. Statements in each thread may or may not be executed simultaneously with statements in other threads. If they cannot be executed (based on scheduling conflicts) they will be interleaved, in accordance with regular rule scheduling. The entire par block terminates when the last of its contained threads terminates.

In the conditional form (if...else), one or the other branch FSM is executed, depending on the boolean condition. The total time taken is the total time for the chosen branch.

Loops within a Stmt context express time-based (temporal) behavior, unlike loops elsewhere in BSV which are statically elaborated. There are two forms, while and for. In both forms, the loop body statements can contain the keywords continue or break, where continue immediately jumps to the start of the next iteration and break jumps out of the loop to the loop sequel. There is also a repeat(n) form where n is a static integer constant.

14.3.2 AutoFSM

Source directory: fsm/autofsm

Source listing: Appendix A.13.4

Often for tests and testbenches, we just want to start up a sequence automatically and let it run once. If we don't need to start and restart it, then we can use mkAutoFSM. Basically, it's like any other FSM except it starts running a few cycles after reset and it calls $finish when its done. Be careful that the work in the rest of your system has completed before the AutoFSM calls $finish.

```
package Tb;

import StmtFSM::*;

(* synthesize *)
module mkTb (Empty);
   Stmt test =
   seq
      $display("I am now running at ", $time);
      $display("I am now running one more step at ", $time);
      $display("And now I will finish at", $time);
   endseq;

   mkAutoFSM ( test );

   rule always-run;
      $display("    and a rule fires at", $time);
   endrule
endmodule
endpackage
```

Another common idiom is to write the `Stmt` expression directly as an argument to `mkAutoFSM`:

```
mkAutoFSM (
    seq
       $display("I am now running at ", $time);
       $display("I am now running one more step at ", $time);
       $display("And now I will finish at", $time);
    endseq;
);
```

14.3.3   StmtFSM example

Source directory: `fsm/example`

Source listing: Appendix A.13.5

Now that we've examined some of the StmtFSM basics, let's look at a larger example. This example doesn't really do anything interesting other than demonstrate many of the components of StmtFSM.

As always, to use the StmtFSM library, we must import the `StmtFSM` package. We also instantiate a `FIFOF` and some registers to introduce some subtleties into the FSM behavior.

```
package Tb;

import StmtFSM::*;
import FIFOF::*;

(* synthesize *)
module mkTb (Empty);
   FIFOF#(int) fifo <- mkFIFOF;
   Reg#(Bool)  cond <- mkReg(True);
   Reg#(int)     ii <- mkRegU;
   Reg#(int)     jj <- mkRegU;
   Reg#(int)     kk <- mkRegU;
```

We define two functions to use in our example. The first, `functionOfAction` is an example of an action inside a function. The second is `functionOfSeq`, is a function which returns a `seq` statement.

```
function Action functionOfAction( int x );
   return action
             $display("an action inside of a function, writing %d", x);
             ii <= x;
          endaction;
endfunction
```

```
function Stmt functionOfSeq( int x );
   return seq
              action
                 $display("sequence in side function %d", x);
                 ii <= x;
              endaction
              action
                 $display("sequence in side function %d", x+1);
                 ii <= x+1;
              endaction
           endseq;
endfunction
```

The name of our FSM spec is test, of type Stmt, which takes a sequence of actions. The function
functionOfAction returns a value of type Action, defining an action that we will be repeating.

```
Stmt test =
seq
   $display("This is action 1"); // cycle 1
   $display("This is action 2"); // cycle 2
   $display("This is action 3"); // cycle 3

   // an action with several actions still only take one cycle
   action
      $display("This is action 4, part a");
      $display("This is action 4, part b");
   endaction

   functionOfAction( 10 );
   functionOfAction( 20 );
   functionOfAction( 30 );
```

We can also create a function containing a sequence (seq...endseq) definition, and call it as an
Action. This is the functionOfSeq we defined above.

```
   functionOfSeq( 50 );

   // this whole action is just one cycle when it fires
   action
      $display("This is action 5");
      if (cond)
         $display("And it may have several actions inside of it");
   endaction
```

A valid action is noAction, basically a *nop*. You can repeat multiple cycles of noAction or of a specific Action. Sequence statements can be nested; our seq...endseq can contain other seq...endseq statements.

```
    repeat (4) noAction;

    repeat (4) action
                    $display("Repeating action!");
                    $display("Check it out");
                endaction;
    seq
       noAction;
       noAction;
    endseq
```

An if/else statement can contain sequences, which will take multiple cycles to execute. The if/else can be inside an action, in which case it takes only one cycle when it executes (either the if or the else branch).

```
    if ( cond ) seq
       // if cond is true, then execute this sequence
       $display("If seq 1");  // cycle 1
       $display("If seq 2");  // cycle 2
       $display("If seq 3");  // cycle 3
    endseq
    else seq
       // if cond is false, then execute this sequence
       action
          $display("Else seq 1");
       endaction
       action
          $display("Else seq 2");
       endaction
    endseq

    // This takes one cycle when it executes
    // The arms can only be of type Action, not Stmt
    action
        if (cond)
           $display("if action 1");
        else
           $display("else action 1");
    endaction
```

Implicit conditions still matter. In the following enq statements, the FSM will stall on the action when the implicit condition is false. In this example, we are forcing this by having a rule pull data out of the FIFO, but only every 4 cycles. This means the FIFO will fill up, and then the FSM will stall until the rule pulls some data out of the FIFO.

```
action
    $display("Enq 10 at time ", $time);
    fifo.enq( 10 );
endaction
action
    $display("Enq 20 at time ", $time);
    fifo.enq( 20 );
endaction
action
    $display("Enq 30 at time ", $time);
    fifo.enq( 30 );
endaction
action
    $display("Enq 40 at time ", $time);
    fifo.enq( 40 );
endaction
```

await is a function provided by the StmtFSM package and is used to generate an implicit condition; await blocks the execution of an action until a condition becomes True. In this case, we want to print a message when we are done. While the FIFO is notEmpty, the state machine waits and no actions fire.

```
action
    $display("FIFO is now empty, continue...");
    await( fifo.notEmpty() == False );
endaction
```

for loops are handy, but notice one small timing quirk. The loop body takes two cycles, one for the compare/first action and a second for the action to increment. This is done so that the increment doesn't interfere with the last step (which, in a large machine, may be complicated). A while loop, on the other hand, has tighter control. The init is one cycle and the while check and action is also a single cycle.

```
for (ii <= 0; ii < 10; ii <= ii + 1) seq
    $display("For loop step %d at time ", ii, $time);
endseq

ii <= 0;                 //init for while loop
while (ii < 10) seq  //check and action is one cycle
    action
        $display("While loop step %d at time ", ii, $time);
        ii <= ii + 1;
    endaction
endseq
```

Loops and control statements can be nested as deep as required.

```
for (ii <=0; ii<=4; ii<=ii+1)
    for (jj <=0; jj<=5; jj<=jj+1)
        for (kk <=0; kk<=3; kk<=kk+1) seq
            action
                if (kk == 0)
                    $display("kk = 000, and ii=%1d, jj=%1d", ii, jj );
                else
                    $display("kk = -%1d, and ii=%1d, jj=%1d", kk, ii, jj );
            endaction
        endseq
```

We can use par/endpar within the seq/endseq block. In a par block, each line is a Stmt or action that is executed in parallel. The entire par block completes when all the sub sequences finish.

```
par
    $display("Par block statement 1 at ", $time);

    $display("Par block statement 2 at ", $time);

    seq
        $display("Par block statement 3a at ", $time);
        $display("Par block statement 3b at ", $time);
    endseq
endpar
$display("Par block done!");
```

The following actions/sequences may stall if there are implicit conditions. Since these statements run in parallel, we expect each sequence to run to completion as it can. Since there can be conflicts with other statements in the sequence, it can be tricky to see exactly what is going to happen.

```
        par
            seq
                ii <= ii + 2;
                ii <= ii + 3;
            endseq

            seq
                ii <= ii + 1;
                ii <= ii + 0;
            endseq

            ii <= ii + 10;
        endpar
        $display("par block conflict test, ii = ", ii);
```

Let's look at the compile-time warnings. For example, the above statements will generate the following warning:

```
Warning: "Tb.bsv", line 36, column 8: (G0036)
  Rule "testFSM_mod_list_1_vs_vs_v_action_l256c16" will appear to fire before
  "testFSM_mod_list_1_vs_v_action_l252c13" when both fire in the same clock
    cycle, affecting:
  calls to ii.write vs. ii.write
```

One very helpful thing to note is the generated rule name has at the end of it a useful piece of information: `testFSM_mod_list_1_vs_vs_v_action_l256c16`, the last characters of the rule names are often the line and character position (l<line>c<char>). In this case, we can see that the warning is talking about line 256, character 16.

This is the end of the seq statement defining the FSM named `test`.

```
    endseq;
```

Now we'll instantiate an FSM interface type with instance name `testFSM`. `mkFSM` takes the `Stmt` as a parameter.

The FSM interface has a few methods defined:

- method Action start: start running state machine, if it's not running already
- method Bool done : return true if FSM is not running. `done` is set to `True` at reset.

```
  FSM testFSM <- mkFSM( test );
```

We'll now define another FSM to empty the FIFO.

```
Stmt rcvr =
seq
   // loop forever, watching for data out of the FIFO
   while(True) seq
      action
         $display("FIFO popped data", fifo.first());
         fifo.deq();
      endaction
      repeat (5) noAction;  // delay 5 cycles between dequeues
   endseq
endseq;

FSM rcvrFSM <- mkFSM( rcvr );
```

The rule `startit` starts the main loop FSM, `testFSM` and the FIFO loop FSM, `rcvrFSM`. The former will eventually terminate; the latter will never terminate because it contains a `while(True)` loop.

```
rule startit;
   testFSM.start();
   rcvrFSM.start();
endrule
```

The rule `startit` does not fire again, since the implicit conditions on the `.start` methods go false as long as the FSMs are running.

We could have put the `$finish` statement at the end of the FSM, but the following rule shows how you could use the `done` method.

```
rule finish (testFSM.done() && !rcvrFSM.done());
   $finish;
endrule

endmodule
endpackage
```

Chapter 15

Importing existing RTL into a BSV design

The BSV `import "BVI"` statement is used to wrap a Verilog module so that it looks like a BSV module and can be used in a BSV design. This allows you to reuse RTL components from previous designs or RTL that has been generated by other tools. With this technique you can also write custom Verilog primitives for use in multiple designs. Many of the BSV primitives (registers, FIFOs, etc), are implemented in Verilog, and then wrapped for inclusion in BSV designs. Instead of Verilog ports, the wrapped module has methods and interfaces.

The Bluespec Development Workstation contains the **importBVI Wizard** to help you write the `import "BVI"` statement. This option is accessed from the **Tools** menu of the workstation. The wizard proceeds through six steps:

1. Verilog Module Overview: Review the Verilog parameters, inputs, outputs, and inouts

2. Bluespec Module Definition: Create the module header for the `import BVI` statement

3. Method Port Binding: Bind methods, ports, and subinterfaces to Verilog inputs and outputs

4. Combinational Paths: Add the `path` statements

5. Scheduling Annotations: Add the `schedule` statements

6. Finish: Review, compile, and save the BSV wrapper.

When you create the wrapper for a Verilog file, you choose how to translate the Verilog inputs and outputs into methods and interfaces. A single Verilog module can have multiple implementations in BSV, each connecting the Verilog pins to different interface methods.

We're going to look at two examples of import statements for a sizedFIFO Verilog file. The first example uses the `FIFOF` interface from the BSV library while the second defines a new interface, based on `Get` and `Put` interfaces.

Both examples are wrapping the Verilog file `SizedFIFO.v`. Note that in these examples the Verilog and RTL signal names are in all CAPS, while the BSV names are in mixed case. The Verilog file can be found in Appendix A.14.1.

```
module SizedFIFO(V_CLK, V_RST_N, V_D_IN, V_ENQ,
                 V_FULL_N, V_D_OUT, V_DEQ, V_EMPTY_N, V_CLR);

   parameter                       V_P1WIDTH = 1; // data width
   parameter                       V_P2DEPTH = 3;
   parameter                       V_P3CNTR_WIDTH = 1; // log(V_P2DEPTH-1)
   // The -1 is allowed since this model has a fast output register
   parameter                       V_GUARDED = 1;

   input                           V_CLK;
   input                           V_RST_N;
   input                           V_CLR;
   input [V_P1WIDTH - 1 : 0]       V_D_IN;
   input                           V_ENQ;
   input                           V_DEQ;

   output                          V_FULL_N;
   output                          V_EMPTY_N;
   output [V_P1WIDTH - 1 : 0]      V_D_OUT;
```

15.1   Using an existing interface

Source directory: `importbvi/FIFOifc`

Source listing: Appendix A.14.2

In this section we're going to wrap the Verilog file `SizedFIFO.v` with a BSV module providing the FIFOF interface. The definition of the FIFOF, from the FIFOF package, is:

```
interface FIFOF #(type a) ;
    method Action enq(a x1) ;
    method Action deq() ;
    method a first() ;
    method Action clear() ;
    method Bool notFull() ;
    method Bool notEmpty() ;
endinterface: FIFOF
```

The `import "BVI"` statement defines a module named `mkSizedFIFO`, which has two arguments, `depth` and `g`, and provides the FIFOF interface. The `method` statements shown below connects the methods in the FIFOF interface to the appropriate Verilog wires. The *inhigh* attribute on an enable indicates that the method is always_enabled. The source code for this example is in the file SizedFIFO1.bsv.

15.1.1 Module header definition

The statement starts with the import "BVI" keyword, followed by the name of the Verilog module being imported. The Verilog name can be excluded if it is the same as the BSV name for the module. This is followed by the usual BSV module definition statement.

```
import "BVI" SizedFIFO =
module mkSizedFIFO #(Integer depth, Bool guard) (FIFOF#(a))
        provisos(Bits#(a,size_a));
```

In our example, the names of the Verilog and BSV modules do not match. But if they did, as they do in the following header, which both use the name RWire, then the Verilog name can be excluded.

```
import "BVI" RWire =
   module RWire (VRWire#(a))
      provisos (Bits#(a,sa));
   ...
   endmodule: vMkRWire
```

could be also written as:

```
import "BVI"
   module RWire (VRWire#(a))
      provisos (Bits#(a,sa));
   ...
   endmodule: vMkRWire
```

15.1.2 Parameters

The parameter statements define the compile-time constants passed into the Verilog module. The valid types for parameters are String, Integer and Bit#(n). Parameter expressions must be compile-time constants. They often include expressions using or containing type information from the interface. Our sizedFIFO example has the following parameters:

```
parameter V_P1WIDTH = valueOf(size_a);
parameter V_P2DEPTH = depth;
parameter V_P3CNTR_WIDTH = log2(depth+1);
parameter V_GUARDED = Bit#(1)'(pack(guard));
```

15.1.3   Clocks and Resets

The following statements are used to connect Verilog clock and reset ports to BSV module clocks and resets, and to associate methods with clock and reset signals:

- `input_clock`

- `default_clock`

- `output_clock`

- `input_reset`

- `default_reset`

- `output_reset`

The `input_clock` and `input_reset` statements specify the port connections for the clocks and resets into the module; they correspond to Verilog input ports. The `output_clock` and `output_reset` statements specify the port connections for clocks and resets provided by the module; they correspond to Verilog outputs. The `default_clock` and `default_reset` statements specify the implicit clock and reset used by methods and submodules without `clocked_by` and `reset_by` statements.

A module with multiple input clocks defines a clock domain crossing; the methods in different domains are clocked by different clocks. These modules allow the safe crossing of signals and data from one clock domain (the source domain) to another (the destination domain).

Some rules to remember when writing `import "BVI"` clock statements:

- Every `import "BVI"` wrapper must specify a default clock and a default reset. They do not have to be connected to Verilog ports and can be defined as `no_clock` or `no_reset`.

- An `input_clock` or `input_reset` statement is required for each RTL input clock or reset port which is not the default.

- For default clocks and resets which are connected to RTL ports, the input and default statements can be combined into a single statement, which is a `default_clock` statement.

- Every `Action` or `ActionValue` method must be associated with a clock. The default clock will be used unless there is a `clocked_by` clause in the method statement. If the default clock is defined as `no_clock`, the method statement must include a `clocked_by` clause specifying an input clock.

Let's look at the clock and reset statements in our example.

```
      default_clock clk;
      default_reset rst_RST_N;

      input_clock clk (V_CLK)  <- exposeCurrentClock;
      input_reset rst_RST_N (V_RST_N) clocked_by(clk)  <- exposeCurrentReset;
```

The first two statements declare the BSV clock named `clk` and the BSV reset named `rest_RST_N` to be the default clock and reset. These are used by methods that don't contain an explicit `clocked_by` or `reset_by` clause. The second set of statements defines the `input_clock` and `input_reset`, connecting them up to the Verilog clock and reset ports. Note that in this example, these are the same clock and reset as the defaults, that is, the Verilog input clock/reset are the default clock/reset for the module. The functions `exposeCurrentClock` and `exposeCurrentReset` return the current clock and reset of the module.

A clock can be connected to two Verilog ports, in which case the first port name within the parentheses is the oscillator, the second port name is the gate. If there is only one port name, it is the oscillator and the gate port is unconnected and defaults to `unused`. Since in this example there is only one Verilog port (`V_CLK`), it is the name of the oscillator and there is no gate port.

15.1.4 Methods

These statements connect the methods defined in the `FIFOF` interface with the inputs and outputs defined in the Verilog file. When defining methods, the Verilog inputs and outputs are translated to signals in the method.

Verilog non-clock inputs translate to:

- Enables for Action or ActionValue methods

- Arguments for any type method - Action, ActionValue or value

Verilog non-clock outputs translate to:

- Return values from ActionValue or value methods

- Ready signals for methods

```
method enq (V_D_IN) enable(V_ENQ) ready(V_FULL_N);
method deq () enable(V_DEQ) ready(V_EMPTY_N);
method V_D_OUT first () ready(V_EMPTY_N);
method V_FULL_N notFull ();
method V_EMPTY_N notEmpty ();
method clear () enable (V_CLR);
```

The following table shows the Verilog parameters, inputs, and outputs along with their BSV translations to the `FIFOF` methods and BSV definitions:

Translation of Verilog into BSV			
SizedFIFO module providing FIFOF interface			
Verilog	BSV		
Statement	Type	Value	Description
`parameter V_P1WIDTH = 1`	parameter	`valueOf(size_a)`	function based on data type
`parameter V_P2DEPTH = 3`	parameter	`depth`	argument to module
`parameter V_P3CNTR_WIDTH = 1`	parameter	`log2(depth+1)`	expression containing parameter
`parameter V_GUARDED=1`	parameter	`Bit#(1)'(pack(g))`	argument to module
`input V_CLK`	Clock	`clk`	default clock
`input V_RST_N`	Reset	`rst_RST_N`	default reset
`input V_CLR`	Action	`clear`	enable for `clear`
`input [V_P1WIDTH - 1:0] D_IN`	Action	`enq`	argument for `enq`
`input V_ENQ`			enable for `enq`
`output V_FULL_N`	Value	`notFull`	ready for `enq` return value of `notFull`
`output [V_P1WIDTH - 1:0] V_D_OUT`	Value	`first`	return value of `first`
`input V_DEQ`	Action	`deq`	enable for `deq`
`output V_EMPTY_N`	Value	`notEmpty`	ready for `deq` ready for `first` return value of `notEmpty`

15.2 Defining a new interface

Source directory: `importbvi/GetPutifc`

Source listing: Appendix A.14.3

In this example, instead of using the pre-defined `FIFOF` interface, we'll define a new interface, `MyGetPut`. The interface has two subinterfaces: a `Get#(a)` interface and a `Put#(a)` interface. The only methods defined will be `get` method and a `put` method. Contrast this with the previous example, in which 6 methods were defined. In this example the Verilog port `V_CLR` is tied off, so there is no `clear` method defined.

The `MyGetPut` interface declaration is:

```
interface MyGetPut#(type t);
   interface Get#(t) g;
   interface Put#(t) p;
endinterface
```

The Verilog parameters, inputs, and outputs are translated in to BSV objects and definitions according to the following table.

Translation of Verilog into BSV			
SizedFIFO module using MyGetPut Interface			
Verilog	BSV		
Statement	Type	Value	Description
`parameter V_P1WIDTH = 1`	parameter	`valueOf(size_a)`	function based on data type
`parameter V_P2DEPTH = 3`	parameter	`depth`	argument to module
`parameter V_P3CNTR_WIDTH = 1`	parameter	`log2(depth+1)`	expression containing parameter
`parameter V_GUARDED=1`	parameter	`Bit#(1)'(pack(guard))`	argument to module
`input V_CLK`	Clock	`clk`	default clock
`input V_RST_N`	Reset	`rst_RST_N`	default reset
`input V_CLR`	port	`0`	tied off
`input [V_P1WIDTH - 1:0] D_IN`	Action	`put`	argument for p_put
`input V_ENQ`			enable for p_put
`output V_FULL_N`			ready for p_put
`output [V_P1WIDTH - 1:0] V_D_OUT`	Value	`g_get`	output value
`input V_DEQ`			enable for g_get
`output V_EMPTY_N`			ready for g_get

The module header provides the interface `MyGetPut` instead of the `FIFOF` interface. We have to import the `GetPut` package because we're using the `Get#(a)` and `Put#(a)` interfaces defined there.

```
import GetPut :: * ;

import "BVI" SizedFIFO =
module mkSizedFIFO #(Integer depth, Bool guard) (MyGetPut#(a))
        provisos(Bits#(a, size_a));
```

The parameters are the same. The clock and reset statements are also the same as the previous example. In this example, we have a `port` statement, tying the `V_CLR` pin to zero.

The `port` statement declares an input port, which is not part of a method, along with the value to be passed to the port. While parameters must be compile-time constants, ports can be dynamic (such as the `_read` method of a register instantiated elsewhere in the module body). The `port` statement is analogous to arguments to a BSV module, but it is rarely needed since the BSV style is to interact and pass arguments through methods. Verilog ports are usually connected to BSV methods. Either of the defining operators `<-` or `=` may be used, as appropriate.

```
      parameter V_P1WIDTH = valueOf(size_a);
       parameter V_P2DEPTH = depth;
       parameter V_P3CNTR_WIDTH = log2(depth+1);
       parameter V_GUARDED = Bit#(1)'(pack(guard));


       port V_CLR = 0 ;


       default_clock clk;
       default_reset rst_RST_N;


       input_clock clk (V_CLK)   <- exposeCurrentClock;
       input_reset rst_RST_N (V_RST_N) clocked_by(clk)   <- exposeCurrentReset;
```

The biggest change is in the definition of the methods. The methods defined in the MyGetPut
interface are different than the methods defined in the FIFOF interface.

```
      interface Get g;
          method V_D_OUT get () enable(V_DEQ) ready(V_EMPTY_N);
      endinterface

      interface Put p;
          method put(V_D_IN) enable(V_ENQ) ready(V_FULL_N);
      endinterface
```

The two examples use the same inputs and outputs, but implement very different interface methods.
The table below compares how the Verilog inputs and outputs are used to define the interface
methods.

Translation of Verilog into BSV		
Verilog Signal	FIFOF interface	MyGetPut interface
V_CLR	enable for clear method	tied off to 0
D_IN	argument for enq	argument for p_put
V_ENQ	enable for enq	enable for p_put
V_FULL_N	ready for enq return value of notFull	ready for p_put
V_D_OUT	output value of first	output value of g_get
V_DEQ	enable for deq	enable for g_get
V_EMPTY_N	ready for deq ready for first return value of notEmpty	ready for g_get

Appendix A

Source Files

Complete, compilable, source and workstation project files are provided for all examples within this manual. This appendix contains the complete source code listing for each of the examples. The BSV (.bsv) and workstation project (.bspec) files are found in the directory listed for each example.

The source code directories follow the organization of the manual; each chapter corresponds to a directory, with each example having its own subdirectory.

For example, the first chapter contains three examples: Hello_world, a testbench communicating with DUT (multiple modules) and an example divided into multiple packages. The directory structure for these examples is:

```
getting_started/hello_world
getting_started/mult_modules
getting_started/mult_packages
```

Within each of these directories is a .bspec file for use with the development workstation, and all .bsv files containing the source code.

A.1 Getting started with BSV

Chapter examples are in directory: getting_started

A.1.1 A simple example

Source files are in directory: getting_started/hello_world

```
// Copyright 2008 Bluespec, Inc.  All rights reserved.
```

```
// ==================================================================
// Small Example Suite: Example 1a
// Very basic 'Hello World' example just to go through the
// mechanics of actually building and running a small BSV program
// (and not intended to teach anything about hardware!).
// ==================================================================

package Tb;

(* synthesize *)
module mkTb (Empty);

   rule greet;
      $display ("Hello World!");
      $finish (0);
   endrule

endmodule: mkTb

endpackage: Tb


// ==================================================================
```

A.1.2   Testbench communicating with DUT

Source files are in directory: getting_started/mult_modules

```
package Tb;

(* synthesize *)
module mkTb (Empty);

   Ifc_type  ifc <- mkModuleDeepThought;

   rule theUltimateAnswer;
      $display ("Hello World! The answer is: %0d",  ifc.the_answer (10, 15, 17));
      $finish (0);
   endrule

endmodule: mkTb

interface Ifc_type;
   method int the_answer (int x, int y, int z);
endinterface: Ifc_type

(* synthesize *)
module mkModuleDeepThought (Ifc_type);

   method int the_answer (int x, int y, int z);
```

```
      return x + y + z;
   endmethod

endmodule: mkModuleDeepThought

endpackage: Tb
```

A.1.3 Multiple packages in a single design

Source files are in directory: `getting_started/mult_packages`

```
// Copyright 2010 Bluespec, Inc.  All rights reserved.
package Tb;

import DeepThought::*;

(* synthesize *)
module mkTb (Empty);

   Ifc_type  ifc <- mkModuleDeepThought;

   rule theUltimateAnswer;
      $display ("Hello World! The answer is: %0d",  ifc.the_answer (10, 15, 17));
      $finish (0);
   endrule

endmodule: mkTb

endpackage: Tb


package DeepThought;

interface Ifc_type;
   method int the_answer (int x, int y, int z);
endinterface: Ifc_type

(* synthesize *)
module mkModuleDeepThought (Ifc_type);

   method int the_answer (int x, int y, int z);
      return x + y + z;
   endmethod

endmodule: mkModuleDeepThought

endpackage: DeepThought
```

A.2   Data types

Chapter examples are in directory: data

A.2.1   Using abstract types instead of Bit

Source files are in directory: data/bit_types

```
// Copyright 2010 Bluespec, Inc.  All rights reserved.
package Tb;


(* synthesize *)
module mkTb (Empty);

   Reg#(int) step <- mkReg(0);

   Reg#(Int#(16))  int16  <- mkReg('h800);
   Reg#(UInt#(16)) uint16 <- mkReg('h800);

   rule step0 ( step == 0 );
      $display("== step 0 ==");

      //    UInt#(16) foo = -1;  // uncomment this to see error
      UInt#(16) foo = 'h1fff;
      $display("foo = %x", foo);

      foo = foo & 5;
      $display("foo = %x", foo);

      //      [] is not legal for UInt (as it is for Bit)
      //      $display("foo[0] = %x", foo[0]); // uncomment this and see error

      foo = 'hffff;
      $display("foo = %x", foo);

      // and it wraps just like you would expect it to
      foo = foo + 1;
      $display("foo = %x", foo);

      // beware the < 0 is never true (this is an unsigned number)
      // in fact, the compiler will display this the same as
      //      $display("fooneg = %x", 1'd0);
      $display("fooneg = %x", foo < 0) ;

      // this is a quick way to get min and max
      UInt#(16) maxUInt16 = unpack('1);  // all 1's
      UInt#(16) minUInt16 = unpack(0);
```

```
   $display("maxUInt16 = %x", maxUInt16);
   $display("minUInt16 = %x", minUInt16);

   $display("%x < %x == %x (unsigned)", minUInt16, maxUInt16, minUInt16 < maxUInt16);

   step <= step + 1;
endrule

////////////////////////////////////////////////////////////////////
rule step1 ( step == 1 );
   $display("== step 1 ==");

   // this is a quick way to get min and max
   Int#(16) maxInt16 = unpack({1'b0,'1});   // 'h011111
   Int#(16) minInt16 = unpack({1'b1,'0});   // 'h100000

   $display("maxInt16 = %x", maxInt16);
   $display("minInt16 = %x", minInt16);

   $display("%x < %x == %x (signed)", minInt16, maxInt16, minInt16 < maxInt16);

   // you can't do bit manipulation directly...
   // uncomment this to see the error generated
   // $display("error1 = %x", minInt16[4]);

   $display("maxInt16/4 = %x", maxInt16 / 4);

   int16 <= int16 / 4;

   step <= step + 1;
endrule

////////////////////////////////////////////////////////////////////
rule step2 ( step == 2 );
   $display("== step 2 ==");

   Int#(32) bar = 10;

   int foo = bar;
   $display("foo is 32 bits = %b", foo);

   bit onebit = 1;
   Bit#(1) anotherbit = onebit;

   $display("this is 1 bit = %b", onebit);

   step <= step + 1;
endrule
```

```
///////////////////////////////////////////////////////////////////
rule step3 ( step == 3 );
   $display("== step 3 ==");

   Bool b1 = True;
   Bool b2 = True;
   Bool b3 = b1 && b2;

   if ( b1 )
      $display("b1 is True");

   bit onebit = 1;

   // if (onebit)                    // uncomment this to see the type error
   //    $display("onebit is a 1");

   if (onebit == 1)
      $display("onebit is a 1");

   step <= step + 1;
endrule

///////////////////////////////////////////////////////////////////
rule step4 ( step == 4 );
   $display ("=== ===");
   $display ("All done");
   $finish (0);
endrule
endmodule: mkTb

endpackage: Tb
```

A.2.2   Integers

Source files are in directory: data/integers

```
// Copyright 2010 Bluespec, Inc.  All rights reserved.
package Tb;

(* synthesize *)
module mkTb (Empty);
   Integer inx = 0;

   Bool arr1[4];
   arr1[inx  ] = True;

   inx = inx + 1;
   arr1[inx] = False;
```

```
    inx = inx + 1;
    arr1[inx] = True;

    inx = inx + 1;
    arr1[inx] = True;

    int arr2[16];
    for (Integer i=0; i<16; i=i+1)
        arr2[i] = fromInteger(i);

    Integer foo = 10;
    foo = foo + 1;
    foo = foo * 5;
    Bit#(16) var1 = fromInteger( foo );

    ///////////////////////////////////////////////////////////////
    Reg#(int) step <- mkReg(0);

    rule init ( step == 0 );
        $display("=== step 0 ===");

        Integer x = 10 ** 10000;
        Bool isLarge = (x > 0);
        $display("isLarge = ", isLarge);

        // uncomment this and see the big number created, causing a error :)
        //$display("Large number = ", x);

        step <= step + 1;
    endrule

    rule step1 ( step == 1 );
        $display("=== step 1 ===");

        for (Integer i=1; i<16; i=i*2)
            $display("i is ", i);

        step <= step + 1;
    endrule

    rule step2 ( step == 2 );
        $display ("All done");
        $finish (0);
    endrule

endmodule: mkTb

endpackage: Tb
```

A.2.3   Strings

Source files are in directory: data/strings

```
// Copyright 2010 Bluespec, Inc.  All rights reserved.
package Tb;

(* synthesize *)
module mkTb (Empty);

   function Action passString( String s );
      return action
              $display("passString => ", s);
              endaction;
   endfunction

   Reg#(int) step <- mkReg(0);

   /////////////////////////////////////////////////////////////////
   rule init ( step == 0 );
      $display("=== step 0 ===");

      String s1 = "This is a test";
      $display("first string = ", s1);

      // we can use + to concatenate
      String s2 = s1 + " of concatenation";
      $display("Second string = ", s2);

      passString( "String passed to a function" );

      // the valid escape chars are listed in the reference manual
      step <= step + 1;
   endrule

   /////////////////////////////////////////////////////////////////
   rule step1 ( step == 1 );
      $display ("All done");
      $finish (0);
   endrule

endmodule: mkTb

endpackage: Tb
```

A.3   Variables, assignments, and combinational circuits

Chapter examples are in directory: `variable`

A.3.1 Variable declaration and initialization

Source files are in directory: `variable/declaration`

```
// Copyright 2010 Bluespec, Inc.  All rights reserved.

package Tb;

import FIFO::*;
import GetPut::*;
import ClientServer::*;

// A struct representing a 2-dimensional 'coordinate'
typedef struct { int x; int y; } Coord
    deriving (Bits);

// An interface, and a module with that interface

module mkTransformer (Server#(Coord, Coord));
   FIFO#(Coord) fi <- mkFIFO;
   FIFO#(Coord) fo <- mkFIFO;

   Coord delta1 = Coord { x: 10, y: 20 };
   let   delta2 = Coord { x: 5,  y: 8 };

   rule transform;
      Coord c = fi.first();  fi.deq();
      c.x = c.x + delta1.x + delta2.x;
      c.y = c.y + delta1.y + delta2.y;
      fo.enq (c);
   endrule

   interface request = toPut (fi);
   interface response = toGet (fo);

endmodule: mkTransformer

// ----------------------------------------------------------------

(* synthesize *)
module mkTb (Empty);

   Reg#(int) cycle <- mkReg (0);

   Reg#(int) rx <- mkReg (0);
   Reg#(int) ry <- mkReg (0);

   Server#(Coord, Coord) s <- mkTransformer;

   rule count_cycles;
```

```
        cycle <= cycle + 1;
        if (cycle > 7) $finish(0);
    endrule

    rule source;
        let c = Coord { x: rx,  y: ry };
        s.request.put (c);
        rx <= rx + 1;
        ry <= ry + 1;
        $display ("%0d: rule source, sending Coord { x: %0d, y: %0d }", cycle, rx, ry);
    endrule

    rule sink;
        let c <- s.response.get ();
        $display ("%0d: rule sink, returned value is Coord { x: %0d, y: %0d }", cycle, c.x, c.y);
    endrule

endmodule: mkTb

endpackage: Tb
```

A.3.2   Combinational circuits

Source files are in directory: variable/combcircuits

```
// Copyright 2010 Bluespec, Inc.  All rights reserved.
package Tb;

import FIFO::*;
import GetPut::*;
import ClientServer::*;

(* synthesize *)
module mkTb (Empty);

    Reg#(int) cycle <- mkReg (0);

    rule count_cycles;
        cycle <= cycle + 1;
        if (cycle > 7) $finish(0);
    endrule

    int x = 10;

    rule r;
        int a = x;
        a = a * a;
        a = a - 5;
```

```
      if (pack(cycle)[0] == 0) a = a + 1;
      else                     a = a + 2;

      if (pack(cycle)[1:0] == 3) a = a + 3;

      for (int k = 20; k < 24; k = k + 1)
         a = a + k;

      $display ("%0d: rule r, a = %0d", cycle, a);
   endrule

endmodule: mkTb

endpackage: Tb
```

A.4 Rules, registers, and FIFOs

Chapter examples are in directory: `rulesregs`

A.4.1 Defining and updating a Register

Source files are in directory: `rulesregs/register`

```
// Copyright 2010 Bluespec, Inc.  All rights reserved.
package Tb;

(* synthesize *)
module mkTb (Empty);

   Reg#(int) x <- mkReg (23);

   rule countup (x < 30);
      int y = x + 1;
      x <= x + 1;
      $display ("x = %0d, y = %0d", x, y);
   endrule

   rule done (x >= 30);
      $finish (0);
   endrule

endmodule: mkTb

endpackage: Tb
```

A.4.2   Composition of rules

Source files are in directory: rulesregs/rules

```
package Tb;

(* synthesize *)
module mkTb (Empty);

   Reg#(int) x1    <- mkReg (10);
   Reg#(int) y1    <- mkReg (100);

   Reg#(int) x2    <- mkReg (10);
   Reg#(int) y2    <- mkReg (100);

   rule r1;
      $display ("r1");
      x1 <= y1 + 1;
      y1 <= x1 + 1;
   endrule

   rule r2a;
      $display ("r2a");
      x2 <= y2 + 1;
   endrule

   rule r2b;
      $display ("r2b");
      y2 <= x2 + 1;
   endrule

//   (* descending_urgency = "r2a, r2b" *)

   rule show_and_eventually_stop;
      $display ("    x1,y1 = %0d,%0d;    x2,y2 = %0d,%0d", x1, y1, x2, y2);
      if (x1 >= 105) $finish (0);
   endrule

endmodule: mkTb

endpackage: Tb
```

A.4.3   Multiple registers in a rigid pipeline

Source files are in directory: rulesregs/reg_pipeline

```
package Tb;

(* synthesize *)
module mkTb (Empty);

   Reg#(int) x     <- mkReg ('h10);
   Pipe_ifc  pipe <- mkPipe;

   rule fill;
      pipe.send (x);
      x <= x + 'h10;
   endrule

   rule drain;
      let y = pipe.receive();
      $display ("     y = %0h", y);
      if (y > 'h80) $finish(0);
   endrule
endmodule

interface Pipe_ifc;
   method Action send (int a);
   method int     receive ();
endinterface

(* synthesize *)
module mkPipe (Pipe_ifc);

   Reg#(int) x1     <- mkRegU;
   Reg#(int) x2     <- mkRegU;
   Reg#(int) x3     <- mkRegU;
   Reg#(int) x4     <- mkRegU;

   rule r1;
      x2 <= x1 + 1;
      x3 <= x2 + 1;
      x4 <= x3 + 1;
   endrule

   rule show;
      $display ("     x1, x2, x3, x4 = %0h, %0h, %0h, %0h", x1, x2, x3, x4);
   endrule

   method Action send (int a);
      x1 <= a;
   endmethod

   method int receive ();
      return x4;
```

```
   endmethod

endmodule: mkPipe

endpackage: Tb
```

A.4.4 Using register syntax shorthand

Source files are in directory: rulesregs/reg_shorthand

```
// Copyright 2010 Bluespec, Inc.  All rights reserved.
package Tb;

(* synthesize *)
module mkTb (Empty);

   Reg#(int) x    <- mkReg ('h10);
   Reg#(int) pipe <- mkPipe;

   rule fill;
      pipe <= x;
      x <= x + 'h10;
   endrule

   rule drain;
      let y = pipe;
      $display ("    y = %0h", y);
      if (y > 'h80) $finish(0);
   endrule
endmodule

(* synthesize *)
module mkPipe (Reg#(int));

   Reg#(int) x1     <- mkRegU;
   Reg#(int) x2     <- mkRegU;
   Reg#(int) x3     <- mkRegU;
   Reg#(int) x4     <- mkRegU;

   rule r1;
      x2 <= x1 + 1;
      x3 <= x2 + 1;
      x4 <= x3 + 1;
   endrule

   rule show;
      $display ("    x1, x2, x3, x4 = %0h, %0h, %0h, %0h", x1, x2, x3, x4);
   endrule
```

```
   method Action _write (int a);
      x1 <= a;
   endmethod

   method int _read ();
      return x4;
   endmethod

endmodule: mkPipe

endpackage: Tb
```

A.4.5   Using valid bits

Source files are in directory: rulesregs/validbits

```
// Copyright 2010 Bluespec, Inc.   All rights reserved.

package Tb;

(* synthesize *)
module mkTb (Empty);

   Reg#(int) x     <- mkReg ('h10);
   Reg#(int) pipe <- mkPipe;

   rule fill;
      pipe <= x;
      x <= x + 'h10;
   endrule

   rule drain;
      let y = pipe;
      $display ("    y = %0h", y);
      if (y > 'h80) $finish(0);
   endrule
endmodule

(* synthesize *)
module mkPipe (Reg#(int));

   Reg#(Bool) valid1 <- mkReg(False);   Reg#(int) x1     <- mkRegU;
   Reg#(Bool) valid2 <- mkReg(False);   Reg#(int) x2     <- mkRegU;
   Reg#(Bool) valid3 <- mkReg(False);   Reg#(int) x3     <- mkRegU;
   Reg#(Bool) valid4 <- mkReg(False);   Reg#(int) x4     <- mkRegU;

   rule r1;
      valid2 <= valid1;     x2 <= x1 + 1;
      valid3 <= valid2;     x3 <= x2 + 1;
```

```
      valid4 <= valid3;     x4 <= x3 + 1;
   endrule

   rule show;
      $write ("    x1, x2, x3, x4 =");
      display_Valid_Value (valid1, x1);
      display_Valid_Value (valid2, x2);
      display_Valid_Value (valid3, x3);
      display_Valid_Value (valid4, x4);
      $display ("");
   endrule

   method Action _write (int a);
      valid1 <= True; x1 <= a;
   endmethod

   method int _read () if (valid4);
      return x4;
   endmethod

endmodule: mkPipe

function Action display_Valid_Value (Bool valid, int value);
   if (valid) $write (" %0h", value);
   else       $write (" Invalid");
endfunction

endpackage: Tb
```

A.4.6   Asynchronous pipeline using FIFOs

Source files are in directory: `rulesregs/fifo_pipeline`

```
// Copyright 2010 Bluespec, Inc.  All rights reserved.

package Tb;

import FIFO::*;

(* synthesize *)
module mkTb (Empty);

   Reg#(int) x     <- mkReg ('h10);
   Pipe_ifc  pipe <- mkPipe;

   rule fill;
      pipe.send (x);
      x <= x + 'h10;
   endrule
```

```
   rule drain;
      let y <- pipe.receive;
      $display ("    y = %0h", y);
      if (y > 'h80) $finish(0);
   endrule
endmodule

interface Pipe_ifc;
   method Action           send   (int x);
   method ActionValue#(int) receive;
endinterface

(* synthesize *)
module mkPipe (Pipe_ifc);

   FIFO#(int) f1    <- mkFIFO;
   FIFO#(int) f2    <- mkFIFO;
   FIFO#(int) f3    <- mkFIFO;
   FIFO#(int) f4    <- mkFIFO;

   rule r2;
      let v1 = f1.first; f1.deq;
      $display (" v1 = %0h", v1);
      f2.enq (v1+1);
   endrule

   rule r3;
      let v2 = f2.first; f2.deq;
      $display (" v2 = %0h", v2);
      f3.enq (v2+1);
   endrule

   rule r4;
      let v3 = f3.first; f3.deq;
      $display (" v3 = %0h", v3);
      f4.enq (v3+1);
   endrule

   method Action send (int a);
      f1.enq (a);
   endmethod

   method ActionValue#(int) receive ();
      let v4 = f4.first;
      $display (" v4 = %0h", v4);
      f4.deq;
      return v4;
   endmethod
```

```
endmodule: mkPipe

endpackage: Tb
```

A.5 Module hierarchy and interfaces

Chapter examples are in directory: `modules`

A.5.1 Module hierarchies

Source files are in directory: `modules/mod_hierarchy`

```
// Copyright 2010 Bluespec, Inc.  All rights reserved.

package Tb;

(* synthesize *)
module mkTb (Empty);
   Empty m3 <- mkM3;
endmodule

// ----------------------------------------------------------------

(* synthesize *)
module mkM3 (Empty);

   Reg#(int) x    <- mkReg (10);
   M1_ifc    m1b <- mkM1 (20);
   M2_ifc    m2  <- mkM2 (30);

   rule justonce;
      $display ("x = %0d, m1b/x = %0d, m2/x = %0d, m2/m1a/x = %0d",
                x, m1b.read_local_x, m2.read_local_x, m2.read_sub_x);
      $finish (0);
   endrule

endmodule

// ----------------------------------------------------------------

interface M2_ifc;
   method int read_local_x ();
   method int read_sub_x ();
endinterface

(* synthesize *)
module mkM2 #(parameter int init_val) (M2_ifc);
```

```
   M1_ifc  m1a <- mkM1 (init_val + 10);
   Reg#(int) x <- mkReg (init_val);

   method int read_local_x ();
      return x;
   endmethod

   method int read_sub_x ();
      return m1a.read_local_x;
   endmethod

endmodule: mkM2

// ----------------------------------------------------------------

interface M1_ifc;
   method int read_local_x ();
endinterface

(* synthesize *)
module mkM1 #(parameter int init_val) (M1_ifc);

   Reg#(int) x <- mkReg (init_val);

   method int read_local_x ();
      return x;
   endmethod

endmodule: mkM1

endpackage: Tb
```

A.5.2   Implicit conditions of methods

Source files are in directory: `modules/implicit_cond`

```
// Copyright 2010 Bluespec, Inc.  All rights reserved.

package Tb;

(* synthesize *)
module mkTb (Empty);
   Dut_ifc dut <- mkDut;

   rule r1;
      let x1 = dut.f1;
      $display ("x1 = %0d", x1);
      if (x1 > 10) $finish (0);
```

```
      endrule

   rule r2;
      let x2 = dut.f2;
      $display ("    x2 = %0d", x2);
      if (x2 > 10) $finish (0);
   endrule

endmodule
// -----------------------------------------------------------------
interface Dut_ifc;
   method int f1 ();
   method int f2 ();
endinterface

(* synthesize *)
module mkDut (Dut_ifc);

   Reg#(int) x    <- mkReg (0);

   rule count;
      x <= x + 1;
   endrule

   method int f1 ();
      return x;
   endmethod

   method int f2 () if (is_even(x));
      return x;
   endmethod

endmodule

function Bool is_even (int v);
   return ((pack (v))[0] == 0);
endfunction
endpackage: Tb
```

A.5.3   ActionValue method

Source files are in directory: `modules/actionvalue`

```
// Copyright 2010 Bluespec, Inc.  All rights reserved.

package Tb;

(* synthesize *)
module mkTb (Empty);
```

```
    Dut_ifc dut <- mkDut;

    rule r1;
        int w <- dut.avmeth(10);
        $display ("w = %0d", w);
        if (w > 50) $finish (0);
    endrule

endmodule

// -----------------------------------------------------------------

interface Dut_ifc;
    method ActionValue#(int) avmeth (int v);
endinterface

(* synthesize *)
module mkDut (Dut_ifc);

    Reg#(int) x <- mkReg (0);

    method ActionValue#(int) avmeth (int v);
        x <= x + v;
        return x;
    endmethod

endmodule

endpackage: Tb
```

A.5.4 ActionValue method with type error

Source files are in directory: `modules/action_value_type_error`

```
// Copyright 2010 Bluespec, Inc.  All rights reserved.

package Tb;

(* synthesize *)
module mkTb (Empty);
    Dut_ifc dut <- mkDut;

    rule r1;
        int w = dut.avmeth(10);          // DELIBERATE ERROR
        $display ("w = %0d", w);
        if (w > 50) $finish (0);
    endrule

endmodule
```

```
// ------------------------------------------------------------------

interface Dut_ifc;
   method ActionValue#(int) avmeth (int v);
endinterface

(* synthesize *)
module mkDut (Dut_ifc);

   Reg#(int) x <- mkReg (0);

   method ActionValue#(int) avmeth (int v);
      x <= x + v;
      return x;
   endmethod

endmodule

endpackage: Tb
```

A.5.5   ActionValue method with error in rule condition

Source files are in directory: modules/action_value_rule_error

```
// Copyright 2010 Bluespec, Inc.  All rights reserved.

package Tb;

(* synthesize *)
module mkTb (Empty);
   Dut_ifc dut <- mkDut;

   rule r1 (dut.avmeth (10) > 0);      // DELIBERATE ERROR
      int w = dut.avmeth(10);
      $display ("w = %0d", w);
      if (w > 50) $finish (0);
   endrule

endmodule

// ------------------------------------------------------------------

interface Dut_ifc;
   method ActionValue#(int) avmeth (int v);
endinterface

(* synthesize *)
module mkDut (Dut_ifc);
```

```
    Reg#(int) x <- mkReg (0);

    method ActionValue#(int) avmeth (int v);
        x <= x + v;
        return x;
    endmethod

endmodule

endpackage: Tb
```

A.5.6 Defining an Action function

Source files are in directory: modules/action_function

```
// Copyright 2010 Bluespec, Inc.  All rights reserved.

package Tb;

(* synthesize *)
module mkTb (Empty);

    Reg#(int) x <- mkReg (0);
    Reg#(int) y <- mkReg (0);

    function Action incr_both (int dx, int dy);
        return
            action
                x <= x + dx;
                y <= y + dy;
            endaction;
    endfunction

    rule r1 (x <= y);
        incr_both (5, 1);
        $display ("(x, y) = (%0d, %0d); r1 fires", x, y);
        if (x > 30) $finish (0);
    endrule

    rule r2 (x > y);
        incr_both (1, 4);
        $display ("(x, y) = (%0d, %0d): r2 fires", x, y);
    endrule

endmodule

endpackage: Tb
```

A.5.7   Nested interfaces

Source files are in directory: `modules/nested_ifc`

```
// Copyright 2010 Bluespec, Inc.  All rights reserved.

package Tb;

import FIFO::*;

// -----------------------------------------------------------------
// A top-level module connecting the stimulus generator to the DUT

(* synthesize *)
module mkTb (Empty);

   Client_int  stimulus_gen <- mkStimulusGen;
   Server_int  dut          <- mkDut;

   // Connect flow of requests from stimulus generator to DUT
   rule connect_reqs;
      let req <- stimulus_gen.get_request.get (); dut.put_request.put (req);
   endrule

   // Connect flow of responses from DUT back to stimulus generator
   rule connect_resps;
      let resp <- dut.get_response.get (); stimulus_gen.put_response.put (resp);
   endrule
endmodule: mkTb

// -----------------------------------------------------------------
// Interface definitions

interface Put_int;
   method Action put (int x);
endinterface

interface Get_int;
   method ActionValue#(int) get ();
endinterface

interface Client_int;
   interface Get_int  get_request;
   interface Put_int  put_response;
endinterface: Client_int

interface Server_int;
   interface Put_int  put_request;
   interface Get_int  get_response;
endinterface: Server_int
```

```
// ------------------------------------------------------------------
// The DUT

(* synthesize *)
module mkDut (Server_int);

   FIFO#(int) f_in  <- mkFIFO;    // to buffer incoming requests
   FIFO#(int) f_out <- mkFIFO;    // to buffer outgoing responses

   rule compute;
      let x = f_in.first; f_in.deq;
      let y = x+1;                 // Some 'server' computation (here: +1)
      if (x == 20) y = y + 1;      // Modeling an 'occasional bug'
      f_out.enq (y);
   endrule

   interface Put_int put_request;
      method Action put (int x);
         f_in.enq (x);
      endmethod
   endinterface

   interface Get_int get_response;
      method ActionValue#(int) get ();
         f_out.deq; return f_out.first;
      endmethod
   endinterface

endmodule: mkDut

// ------------------------------------------------------------------
// The stimulus generator

(* synthesize *)
module mkStimulusGen (Client_int);

   Reg#(int)  x           <- mkReg (0);        // Seed for stimulus
   FIFO#(int) f_out       <- mkFIFO;           // To buffer outgoing requests
   FIFO#(int) f_in        <- mkFIFO;           // To buffer incoming responses
   FIFO#(int) f_expected <- mkFIFO;            // To buffer expected responses

   rule gen_stimulus;
      f_out.enq (x);
      x <= x + 10;
      f_expected.enq (x+1);    // mimic functionality of the Dut 'server'
   endrule

   rule check_results;
      let y     = f_in.first; f_in.deq;
```

```
      let y_exp = f_expected.first; f_expected.deq;
      $write ("(y, y_expected) = (%0d, %0d)", y, y_exp);
      if (y == y_exp) $display (": PASSED");
      else            $display (": FAILED");
      if (y_exp > 50) $finish(0);
   endrule

   interface get_request = interface Get_int;
                              method ActionValue#(int) get ();
                                 f_out.deq; return f_out.first;
                              endmethod
                           endinterface;

   interface Put_int put_response;
      method Action put (int y) = f_in.enq (y);
   endinterface

endmodule: mkStimulusGen

endpackage: Tb
```

A.5.8   Standard connectivity interfaces

Source files are in directory: `modules/connect_ifc`

```
// Copyright 2010 Bluespec, Inc.  All rights reserved.
package Tb;

import FIFO::*;
import GetPut::*;
import ClientServer::*;
import Connectable::*;

// ----------------------------------------------------------------
// A top-level module connecting the stimulus generator to the DUT

(* synthesize *)
module mkTb (Empty);

   Client#(int,int) stimulus_gen <- mkStimulusGen;
   Server#(int,int) dut          <- mkDut;

   // Connect a client to a server (flow of requests and flow of responses)
   mkConnection (stimulus_gen, dut);

endmodule: mkTb

// ----------------------------------------------------------------
// The DUT
```

```
(* synthesize *)
module mkDut (Server#(int,int));

   FIFO#(int) f_in  <- mkFIFO;    // to buffer incoming requests
   FIFO#(int) f_out <- mkFIFO;    // to buffer outgoing responses

   rule compute;
      let x = f_in.first; f_in.deq;
      let y = x+1;                 // Some 'server' computation (here: +1)
      if (x == 20) y = y + 1;      // Modeling an 'occasional bug'
      f_out.enq (y);
   endrule

   interface request  = toPut (f_in);
   interface response = toGet (f_out);

endmodule: mkDut

// ----------------------------------------------------------------
// The stimulus generator

(* synthesize *)
module mkStimulusGen (Client#(int,int));

   Reg#(int)  x            <- mkReg (0);      // Seed for stimulus
   FIFO#(int) f_out        <- mkFIFO;         // To buffer outgoing requests
   FIFO#(int) f_in         <- mkFIFO;         // To buffer incoming responses
   FIFO#(int) f_expected   <- mkFIFO;         // To buffer expected responses

   rule gen_stimulus;
      f_out.enq (x);
      x <= x + 10;
      f_expected.enq (x+1);    // mimic functionality of the Dut 'server'
   endrule

   rule check_results;
      let y     = f_in.first; f_in.deq;
      let y_exp = f_expected.first; f_expected.deq;
      $write ("(y, y_expected) = (%0d, %0d)", y, y_exp);
      if (y == y_exp) $display (": PASSED");
      else            $display (": FAILED");
      if (y_exp > 50) $finish(0);
   endrule

   return fifosToClient (f_out, f_in);

endmodule: mkStimulusGen

// ----------------
```

```
// An interface transformer (a function with interface arguments and results)

function Client#(req_t, resp_t) fifosToClient (FIFO#(req_t) f_reqs, FIFO#(resp_t) f_resps);
   return interface Client
              interface Get request  = toGet (f_reqs);
              interface Put response = toPut (f_resps);
          endinterface;
endfunction: fifosToClient

endpackage: Tb
```

A.6   Scheduling

Chapter examples are in directory: `schedule`

A.6.1   Scheduling error due to parallel composition

Source files are in directory: `schedule/parallel_error`

```
package Tb;

import FIFO ::*;


// (* synthesize *)
// module mkTb1 (Empty);

//    FIFO#(int) f <- mkFIFO;    // Instantiate a fifo

//    // ----------------
//    // Step 'state' from 1 to 10

//    Reg#(int) state <- mkReg (0);

//    rule step_state;
//       if (state > 9) $finish (0);
//       state <= state + 1;
//    endrule

//    // ----------------
//    // Enqueue two values to the fifo from the same rule

//    rule enq (state < 7);
//       f.enq(state);
//       f.enq(state+1);
```

```
//        $display("fifo enq: %d, %d", state, state+1);
//     endrule

// endmodule: mkTb1

// -----------------------------------------------------------------
// A top-level module instantiating a fifo

(* synthesize *)
module mkTb (Empty);

   FIFO#(int) f <- mkFIFO;     // Instantiate a fifo

   // ----------------
   // Step 'state' from 1 to 10

   Reg#(int) state <- mkReg (0);

   rule step_state;
      if (state > 9) $finish (0);
      state <= state + 1;

   endrule

   // ----------------
   // Enqueue two values to the fifo from seperate rules in the testbench
   //

   rule enq1 (state < 7);
      f.enq(state);
      $display("fifo enq1: %d", state);
   endrule

   rule enq2 (state > 4);
      f.enq(state+1);
      $display("fifo enq2: %d", state+1);
   endrule

   // ----------------
   // Dequeue from the fifo, and check the first value
   //

   rule deq;
      f.deq();
      $display("fifo deq: %d", f.first() );
   endrule

endmodule: mkTb

endpackage: Tb
```

A.6.2　Prioritization using descending_urgency

Source files are in directory: schedule/desc_urgency

```
package Tb;

// importing the FIFO library package
import FIFO ::*;

// ----------------------------------------------------------------
// A top-level module instantiating a fifo

(* synthesize *)
module mkTb (Empty);

   FIFO#(int) f <- mkFIFO;     // Instantiate a fifo

   // ----------------
   // Step 'state' from 1 to 10

   Reg#(int) state <- mkReg (0);

   rule step_state;
      if (state > 9) $finish (0);
      state <= state + 1;
   endrule

   // ----------------
   // Enqueue two values to the fifo from seperate rules in the testbench
   //

   (* descending_urgency = "enq1, enq2"*)
   rule enq1 (state < 7);
      f.enq(state);
      $display("fifo enq1: %d", state);
   endrule

   rule enq2 (state > 4);
      f.enq(state+1);
      $display("fifo enq2: %d", state+1);
   endrule

   // ----------------
   // Dequeue from the fifo, and check the first value
   //

   rule deq;
       f.deq();
```

```
         $display("fifo deq : %d", f.first() );
      endrule

endmodule: mkTb

endpackage: Tb
```

A.6.3   descending urgency vs. execution order

Source files are in directory: schedule/exec_order

```
// Copyright 2010 Bluespec, Inc.  All rights reserved.

package Tb;

import FIFO::*;
import GetPut::*;
import ClientServer::*;

(* synthesize *)
module mkTb (Empty);
   Reg#(int) cycle <- mkReg(0);

   Server#(int, int) g <- mkGadget;

   // Send data to gadget every 5 out of 8 cycles
   rule source;
      cycle <= cycle + 1;
      if (pack (cycle)[2:0] < 5) g.request.put (cycle);
   endrule

   rule drain;
      let x <- g.response.get ();
      $display ("%0d: draining %d", cycle, x);
      if (cycle > 20) $finish(0);
   endrule
endmodule: mkTb

// =================================================================
// This module enqueues something on the outfifo whenever it isn't full,
// either a value from infifo (if there is one), or a bubble (otherwise).  It
// also keeps a count of the maximum interval between enqueueing an item and
// enqueuing a bubble (with no intervening item).

// Let us assume that enqueuing an actual available item is more urgent than
// enqueueing a bubble (and so is explicitly specified).

(* synthesize *)
module mkGadget (Server#(int,int));
```

```
   int bubble_value = 42;

   FIFO#(int) infifo <- mkFIFO;
   FIFO#(int) outfifo <- mkFIFO;

   Reg#(int) bubble_cycles <- mkReg(0);
   Reg#(int) max_bubble_cycles <- mkReg(0);

   (* descending_urgency="enqueue_item, enqueue_bubble" *)
   rule enqueue_item;
      outfifo.enq(infifo.first);
      infifo.deq;
      bubble_cycles <= 0;
   endrule

   rule inc_bubble_cycles;
      bubble_cycles <= bubble_cycles + 1;
   endrule

   rule enqueue_bubble;
      outfifo.enq(bubble_value);
      max_bubble_cycles <= max(max_bubble_cycles, bubble_cycles);
   endrule

   interface request  = fifoToPut (infifo);
   interface response = fifoToGet (outfifo);

endmodule: mkGadget

endpackage: Tb
```

A.6.4   mutually_exclusive

Source files are in directory: schedule/mutually_exclusive

```
// Copyright 2010 Bluespec, Inc.  All rights reserved.

package Tb;

(* synthesize *)
module mkTb(Empty);

   // a 3-bit reg with one bit always set
   Reg#(Bit#(3)) x <- mkReg (1);

   Reg#(int)    cycle <- mkReg (0);

   // A register used to induce a conflict
   Reg#(int) y <- mkReg (0);
```

```
   rule rx;
      cycle <= cycle + 1;
      x <= { x[1:0], x[2] };   // rotate the bits
      $display ("%0d: Rule rx", cycle);
   endrule

// (* mutually_exclusive = "ry0, ry1, ry2" *)    // XXX

   rule ry0 (x[0] == 1);
      y <= y + 1;
      $display ("%0d: Rule ry0", cycle);
   endrule

   rule ry1 (x[1] == 1);
      y <= y + 2;
      $display ("%0d: Rule ry1", cycle);
   endrule

   rule ry2 (x[2] == 1);
      y <= y + 3;
      $display ("%0d: Rule ry2", cycle);
   endrule

   rule done (cycle >= 10);
      $finish (0);
   endrule

endmodule: mkTb

endpackage: Tb

/*
 Compile this program twice,
 - once with the line XXX above commented out
     [ generates conflict warning message and priority logic ]
 - once with the line XXX above uncommented
     [ generates no conflict warning message and no priority logic ]

In each case, examine the Verilog code to see the difference the logic.

* ================================================================
*/
```

A.6.5   conflict_free

Source files are in directory: schedule/conflict_free

```
package Tb;

(* synthesize *)
module mkTb(Empty);

   // a 3-bit reg with one bit always set
   Reg#(Bit#(3)) x <- mkReg (1);

   Reg#(int)      cycle <- mkReg (0);

   // A register used to induce a conflict
   Reg#(int) y <- mkReg (0);

   rule rx;
      cycle <= cycle + 1;
      x <= { x[1:0], x[2] };  // rotate the bits
      $display ("%0d: Rule rx", cycle);
   endrule

// (* conflict_free = "ry0, ry1, ry2" *)    // XXX

   rule ry0;
      if (x[0] == 1) y <= y + 1;
      $display ("%0d: Rule ry0", cycle);
   endrule

   rule ry1;
      if (x[1] == 1) y <= y + 2;
      $display ("%0d: Rule ry1", cycle);
   endrule

   rule ry2;
      if (x[2] == 1) y <= y + 3;
      $display ("%0d: Rule ry2", cycle);
   endrule

   rule done;
      if (cycle >= 10) $finish (0);
   endrule

endmodule: mkTb

endpackage: Tb
```

A.6.6   preempts

Source files are in directory: schedule/preempts

```
// Copyright 2010 Bluespec, Inc.  All rights reserved.

package Tb;

import FIFO::*;
import GetPut::*;
import ClientServer::*;

(* synthesize *)
module mkTb (Empty);
   Reg#(int) cycle <- mkReg(0);

   Server#(int, int) g <- mkGadget;

   // Send data to gadget every 5 out of 8 cycles
   rule source;
      cycle <= cycle + 1;
      if (pack (cycle)[2:0] < 5) g.request.put (cycle);
   endrule

   rule drain;
      let x <- g.response.get ();
      $display ("%0d: draining %d", cycle, x);
      if (cycle > 20) $finish(0);
   endrule
endmodule: mkTb

// =================================================================
// This module transfers an item from the infifo to the outfifo whenever
// possible.  The other rule counts the idle cycles.

(* synthesize *)
module mkGadget (Server#(int,int));
   FIFO#(int) infifo <- mkFIFO;
   FIFO#(int) outfifo <- mkFIFO;

   Reg#(int) idle_cycles <- mkReg(0);

   (* preempts="enqueue_item, count_idle_cycles" *)
   rule enqueue_item;
      outfifo.enq(infifo.first);
      infifo.deq;
   endrule

   rule count_idle_cycles;
      idle_cycles <= idle_cycles + 1;
      $display ("Idle cycle %0d", idle_cycles + 1);
   endrule

   interface request  = fifoToPut (infifo);
```

```
   interface response = fifoToGet (outfifo);

endmodule: mkGadget

endpackage: Tb
```

A.6.7   fire_when_enabled, no_implicit_conditions

Source files are in directory: schedule/fire_when_enabled

```
// Copyright 2010 Bluespec, Inc.  All rights reserved.

package Tb;

import GetPut::*;

(* synthesize *)
module mkTb (Empty);

   Reg#(int) cycle <- mkReg (0);
   Reg#(int) x     <- mkReg (0);

   Put#(int) dut <- mkDut ();

   // ----------------

   rule count_cycles;
      cycle <= cycle + 1;
      if (cycle > 15) $finish(0);
   endrule

//    (* descending_urgency = "r2, r1" *)
   (* descending_urgency = "r1, r2" *)

   (* fire_when_enabled, no_implicit_conditions *)
   rule r1 (pack (cycle)[2:0] != 7);
      dut.put (x);
      x <= x + 1;
   endrule

   rule r2;
      x <= x + 2;
   endrule

endmodule: mkTb


// ================================================================

(* synthesize *)
```

```
module mkDut (Put#(int));
   Reg#(int) y <- mkReg (0);

   method Action put (int z);// if (y > 0);
      y <= z;
   endmethod
endmodule

// ================================================================

endpackage: Tb
```

A.6.8   always_ready, always_enabled

Source files are in directory: schedule/always_ready

```
package Tb;

// ================================================================

import PipelinedSyncROM_model::*;

// A synthesizable instance of the polymorphic pipelined sync ROM

typedef UInt#(16) Addr;
typedef UInt#(24) Data;

(* synthesize *)
(* always_ready = "request, response" *)
(* always_enabled = "request" *)

module mkPipelinedSyncROM_model_specific (PipelinedSyncROM#(Addr,Data));
   let m <- mkPipelinedSyncROM_model ("mem_init.data", 0, 31);
   return m;
endmodule

// ----------------------------------------------------------------
// A top-level module that just reads some data from the pipelined ROM

(* synthesize *)
module mkTb (Empty);

   PipelinedSyncROM#(Addr, Data) rom <- mkPipelinedSyncROM_model_specific ();

   Reg#(Addr) ra <- mkReg (0);
   Reg#(int) cycle <- mkReg (0);
```

```
// ----------------
// Step 'address' and feed to the ROM

rule step_addr;
   rom.request (ra);
   if (ra > 15) $finish (0);
   ra <= ra + 1;
   cycle <= cycle + 1;
   $display ("%0d: Sending address %0h", cycle, ra);
endrule

rule show_data;
   $display ("%0d: Data returned is %0h", cycle, rom.response());
endrule

endmodule: mkTb

endpackage: Tb
```

A.6.9   aggressive_conditions

Source files are in directory: schedule/aggressive_conditions

```
// Copyright 2010 Bluespec, Inc.  All rights reserved.

package Tb;

import FIFO::*;

(* synthesize *)
module mkTb (Empty);

   Reg#(int) cycle <- mkReg (0);

   FIFO#(int) f0 <- mkSizedFIFO (3);
   FIFO#(int) f1 <- mkSizedFIFO (3);

   // ----------------
   // RULES

   rule count_cycles;
      cycle <= cycle + 1;
      if (cycle > 15) $finish(0);
   endrule

   rule rA;
      if (pack (cycle)[0] == 0) begin
```

```
            f0.enq (cycle);
            $display ("%0d: Enqueuing %d into fifo f0", cycle, cycle);
         end
         else begin
            f1.enq (cycle);
            $display ("%0d: Enqueuing %d into fifo f1", cycle, cycle);
         end
      endrule

      rule rB;
         f0.deq ();
      endrule

endmodule: mkTb

endpackage: Tb
```

A.6.10   Separating ActionValue method

Source files are in directory: schedule/separate_av

```
// Copyright 2010 Bluespec, Inc.  All rights reserved.

package Tb;

import FIFO::*;

interface Foo_ifc;
   method ActionValue#(int) avx ();
   method int vy ();
   method Action ay ();

   method int get_cycle_count ();
endinterface

(* synthesize *)
module mkFoo (Foo_ifc);

   Reg#(int) x <- mkReg (0);
   Reg#(int) y <- mkReg (0);
   Reg#(int) cycle <- mkReg (0);

   // ----------------
   // RULES

   rule count_cycles;
      cycle <= cycle + 1;
      if (cycle > 3) $finish (0);
```

```
    endrule

    rule incx;
        x <= x + 1;
        $display ("%0d: rule incx; new x = %0d", cycle, x+1);
    endrule

    rule incy;
        y <= y + 1;
        $display ("%0d: rule incy; new y = %0d", cycle, y+1);
    endrule

    // ----------------
    // METHODS

    method ActionValue#(int) avx ();
        $display ("%0d: method avx; setting x to 42; returning old x = %0d", cycle, x);
        x <= 42;
        return x;
    endmethod

    method int vy ();
        return y;
    endmethod

    method Action ay ();
        $display ("%0d: method ay; setting y to 42", cycle);
        y <= 42;
    endmethod

    method int get_cycle_count ();
        return cycle;
    endmethod
endmodule

// ----------------------------------------------------------------

(* synthesize *)
module mkTb (Empty);

    Foo_ifc foo <- mkFoo;

    let cycle = foo.get_cycle_count ();

    // ----------------
    // RULES

    rule rAVX;
        let x <- foo.avx ();
        $display ("%0d: rule rAVX, x = %0d", cycle, x);
```

```
    endrule

    rule rAY;
        $display ("%0d: rule rAY", cycle);
        foo.ay ();
    endrule

    rule rVY;
        let y = foo.vy ();
        $display ("%0d: rule rVY, y = %0d", cycle, y);
    endrule

endmodule: mkTb

endpackage: Tb
```

A.7 RWires and Wire types

Chapter examples are in directory: `rwire`

A.7.1 Simple RWire example

Source files are in directory: `rwire/simplerwire`

```
// Copyright 2010 Bluespec, Inc.  All rights reserved.

package Tb;

// ----------------------------------------------------------------
// A top-level module connecting the stimulus generator to the DUT

(* synthesize *)
module mkTb (Empty);

    Counter c1 <- mkCounter_v1;    // Instantiate a version 1 counter
    Counter c2 <- mkCounter_v2;    // Instantiate a version 2 counter (same interface type)

    // ----------------
    // Step 'state' from 1 to 10 and show both counter values in each state

    Reg#(int) state <- mkReg (0);

    rule step_state;
        if (state > 9) $finish (0);
        state <= state + 1;
    endrule
```

```
    rule show;
        $display ("In state %0d, counter values are %0d, %0d", state, c1.read (), c2.read ());
    endrule

    // ----------------
    // For counter c1, increment by 5 in states 0..6, decrement by 2 in states 4..10
    // The interesting states are 4..6

    rule incr1 (state < 7);
        c1.increment (5);
    endrule

    rule decr1 (state > 3);
        c1.decrement (2);
    endrule

    // ----------------
    // Same for counter c2 (increment by 5 in states 0..6, decrement by 2 in states 4..10)
    // The interesting states are 4..6

    rule incr2 (state < 7);
        c2.increment (5);
    endrule

    rule decr2 (state > 3);
        c2.decrement (2);
    endrule

endmodule: mkTb

// ------------------------------------------------------------------
// Interface to an up-down counter

interface Counter;
    method int read();                        // Read the counter's value
    method Action increment (int di);         // Step the counter up by di
    method Action decrement (int dd);         // Step the counter down by dd
endinterface: Counter

// ------------------------------------------------------------------
// Version 1 of the counter

(* synthesize *)
module mkCounter_v1 (Counter);
    Reg#(int) value1 <- mkReg(0);  // holding the counter's value

    method int read();
        return value1;
    endmethod
```

```
    method Action increment (int di);
       value1 <= value1 + di;
    endmethod

    method Action decrement (int dd);
       value1 <= value1 - dd;
    endmethod

endmodule: mkCounter_v1

// ----------------------------------------------------------------
// Version 2 of the counter

(* synthesize *)
module mkCounter_v2 (Counter);
    Reg#(int) value2 <- mkReg(0);  // holding the counter's value

    RWire#(int) rw_incr <- mkRWire();  // Signal that increment method is being invoked
    RWire#(int) rw_decr <- mkRWire();  // Signal that decrement method is being invoked

    // This rule does all the work, depending on whether the increment or the decrement
    // methods, both, or neither, is being invoked
    (* fire_when_enabled, no_implicit_conditions *)
    rule doit;
       Maybe#(int) mbi = rw_incr.wget();
       Maybe#(int) mbd = rw_decr.wget();
       int    di        = fromMaybe (?, mbi);
       int    dd        = fromMaybe (?, mbd);
       if     ((! isValid (mbi)) && (! isValid (mbd)))
          noAction;
       else if (   isValid (mbi)  && (! isValid (mbd)))
          value2 <= value2 + di;
       else if ((! isValid (mbi)) &&    isValid (mbd))
          value2 <= value2 - dd;
       else // (   isValid (mbi)  &&    isValid (mbd))
          value2 <= value2 + di - dd;
    endrule

    method int read();
       return value2;
    endmethod

    method Action increment (int di);
       rw_incr.wset (di);
    endmethod

    method Action decrement (int dd);
       rw_decr.wset (dd);
    endmethod
```

```
endmodule: mkCounter_v2

endpackage: Tb

/*
-----------------------------------------------------------------
EXERCISE:

(1) Merge the two mkTb rules 'step_state' and 'show' into one rule;
    observe what happens, and explain it. [Hint: compile-time
    scheduling error.]

* ================================================================
*/
```

A.7.2   RWire with pattern matching

Source files are in directory: rwire/pattern_matching

```
// Copyright 2010 Bluespec, Inc.  All rights reserved.

package Tb;

// ----------------------------------------------------------------
// A top-level module connecting the stimulus generator to the DUT

(* synthesize *)
module mkTb (Empty);

   Counter c1 <- mkCounter_v1;    // Instantiate a version 1 counter
   Counter c2 <- mkCounter_v2;    // Instantiate a version 2 counter (same interface type)

   // ----------------
   // Step 'state' from 1 to 10 and show both counter values in each state

   Reg#(int) state <- mkReg (0);

   rule step_state;
      if (state > 9) $finish (0);
      state <= state + 1;
   endrule

   rule show;
      $display ("In state %0d, counter values are %0d, %0d", state, c1.read (), c2.read ());
   endrule

   // ----------------
   // For counter c1, increment by 5 in states 0..6, decrement by 2 in states 4..10
   // The interesting states are 4..6
```

```
   rule incr1 (state < 7);
      c1.increment (5);
   endrule

   rule decr1 (state > 3);
      c1.decrement (2);
   endrule

   // ---------------
   // Same for counter c2 (increment by 5 in states 0..6, decrement by 2 in states 4..10)
   // The interesting states are 4..6

   rule incr2 (state < 7);
      c2.increment (5);
   endrule

   rule decr2 (state > 3);
      c2.decrement (2);
   endrule

endmodule: mkTb

// ----------------------------------------------------------------
// Interface to an up-down counter

interface Counter;
   method int read();                    // Read the counter's value
   method Action increment (int di);     // Step the counter up by di
   method Action decrement (int dd);     // Step the counter down by dd
endinterface: Counter

// ----------------------------------------------------------------
// Version 1 of the counter

(* synthesize *)
module mkCounter_v1 (Counter);
   Reg#(int) value1 <- mkReg(0);  // holding the counter's value

   method int read();
      return value1;
   endmethod

   method Action increment (int di);
      value1 <= value1 + di;
   endmethod

   method Action decrement (int dd);
      value1 <= value1 - dd;
   endmethod
```

```
endmodule: mkCounter_v1

// ------------------------------------------------------------------
// Version 2 of the counter

(* synthesize *)
module mkCounter_v2 (Counter);
   Reg#(int) value2 <- mkReg(0);  // holding the counter's value

   RWire#(int) rw_incr <- mkRWire();  // Signal that increment method is being invoked
   RWire#(int) rw_decr <- mkRWire();  // Signal that decrement method is being invoked

   // This rule does all the work, depending on whether the increment or the decrement
   // methods, both, or neither, is being invoked
   (* fire_when_enabled, no_implicit_conditions *)
   rule doit;
      case (tuple2 (rw_incr.wget(), rw_decr.wget())) matches
         { tagged Invalid,   tagged Invalid  } : noAction;
         { tagged Valid .di, tagged Invalid  } : value2 <= value2 + di;
         { tagged Invalid,   tagged Valid .dd } : value2 <= value2 - dd;
         { tagged Valid .di, tagged Valid .dd } : value2 <= value2 + di - dd;
      endcase
   endrule

   method int read();
      return value2;
   endmethod

   method Action increment (int di);
      rw_incr.wset (di);
   endmethod

   method Action decrement (int dd);
      rw_decr.wset (dd);
   endmethod

endmodule: mkCounter_v2

endpackage: Tb
```

A.7.3   Wire

Source files are in directory: rwire/wire

```
package Tb;
```

```
// ------------------------------------------------------------------
// A top-level module connecting the stimulus generator to the DUT

(* synthesize *)
module mkTb (Empty);

    Counter c1 <- mkCounter_v1;    // Instantiate a version 1 counter
    Counter c2 <- mkCounter_v2;    // Instantiate a version 2 counter (same interface type)

    // ----------------
    // Step 'state' from 1 to 10 and show both counter values in each state

    Reg#(int) state <- mkReg (0);

    rule step_state;
        if (state > 9) $finish (0);
        state <= state + 1;
    endrule

    rule show;
        $display ("In state %0d, counter values are %0d, %0d", state, c1.read (), c2.read ());
    endrule

    // ----------------
    // For counter c1, increment by 5 in states 0..6, decrement by 2 in states 4..10
    // The interesting states are 4..6

    rule incr1 (state < 7);
        c1.increment (5);
    endrule

    rule decr1 (state > 3);
        c1.decrement (2);
    endrule

    // ----------------
    // Same for counter c2 (increment by 5 in states 0..6, decrement by 2 in states 4..10)
    // The interesting states are 4..6

    rule incr2 (state < 7);
        c2.increment (5);
    endrule

    rule decr2 (state > 3);
        c2.decrement (2);
    endrule

endmodule: mkTb

// ------------------------------------------------------------------
```

```
// Interface to an up-down counter

interface Counter;
   method int read();                    // Read the counter's value
   method Action increment (int di);     // Step the counter up by di
   method Action decrement (int dd);     // Step the counter down by dd
endinterface: Counter

// ----------------------------------------------------------------
// Version 1 of the counter

(* synthesize *)
module mkCounter_v1 (Counter);
   Reg#(int) value1 <- mkReg(0);  // holding the counter's value

   method int read();
      return value1;
   endmethod

   method Action increment (int di);
      value1 <= value1 + di;
   endmethod

   method Action decrement (int dd);
      value1 <= value1 - dd;
   endmethod

endmodule: mkCounter_v1

// ----------------------------------------------------------------
// Version 2 of the counter

(* synthesize *)
module mkCounter_v2 (Counter);
   Reg#(int) value2 <- mkReg(0);  // holding the counter's value

   Wire#(int) w_incr <- mkWire();  // Signal that increment method is being invoked
   Wire#(int) w_decr <- mkWire();  // Signal that decrement method is being invoked

   // ---------------- RULES

   (* descending_urgency = "r_incr_and_decr, r_incr_only, r_decr_only" *)

   rule r_incr_and_decr;
      value2 <= value2 + w_incr - w_decr;
   endrule

   rule r_incr_only;
      value2 <= value2 + w_incr;
   endrule
```

```
   rule r_decr_only;
      value2 <= value2 - w_decr;
   endrule

   // --------------- METHODS

   method int read();
      return value2;
   endmethod

   method Action increment (int di);
      w_incr <= di;
   endmethod

   method Action decrement (int dd);
      w_decr <= dd;
   endmethod

endmodule: mkCounter_v2

endpackage: Tb
```

A.7.4 DWire

Source files are in directory: rwire/dwire

```
// Copyright 2010 Bluespec, Inc.  All rights reserved.

package Tb;

// -----------------------------------------------------------------
// A top-level module connecting the stimulus generator to the DUT

(* synthesize *)
module mkTb (Empty);

   Counter c1 <- mkCounter_v1;    // Instantiate a version 1 counter
   Counter c2 <- mkCounter_v2;    // Instantiate a version 2 counter (same interface type)

   // ---------------
   // Step 'state' from 1 to 10 and show both counter values in each state

   Reg#(int) state <- mkReg (0);

   rule step_state;
      if (state > 9) $finish (0);
      state <= state + 1;
   endrule
```

```
    rule show;
        $display ("In state %0d, counter values are %0d, %0d", state, c1.read (), c2.read ());
    endrule

    // ----------------
    // For counter c1, increment by 5 in states 0..6, decrement by 2 in states 4..10
    // The interesting states are 4..6

    rule incr1 (state < 7);
        c1.increment (5);
    endrule

    rule decr1 (state > 3);
        c1.decrement (2);
    endrule

    // ----------------
    // Same for counter c2 (increment by 5 in states 0..6, decrement by 2 in states 4..10)
    // The interesting states are 4..6

    rule incr2 (state < 7);
        c2.increment (5);
    endrule

    rule decr2 (state > 3);
        c2.decrement (2);
    endrule

endmodule: mkTb

// -----------------------------------------------------------------
// Interface to an up-down counter

interface Counter;
    method int read();                      // Read the counter's value
    method Action increment (int di);       // Step the counter up by di
    method Action decrement (int dd);       // Step the counter down by dd
endinterface: Counter

// -----------------------------------------------------------------
// Version 1 of the counter

(* synthesize *)
module mkCounter_v1 (Counter);
    Reg#(int) value1 <- mkReg(0);  // holding the counter's value

    method int read();
        return value1;
    endmethod
```

```
   method Action increment (int di);
      value1 <= value1 + di;
   endmethod

   method Action decrement (int dd);
      value1 <= value1 - dd;
   endmethod

endmodule: mkCounter_v1

// ----------------------------------------------------------------
// Version 2 of the counter

(* synthesize *)
module mkCounter_v2 (Counter);
   Reg#(int) value2 <- mkReg(0);  // holding the counter's value

   Wire#(int) w_incr <- mkDWire (0);  // Signal that increment method is being invoked
   Wire#(int) w_decr <- mkDWire (0);  // Signal that decrement method is being invoked

   // ---------------- RULES

   // This rule does all the work.
   // If the increment method is being invoked,
   //        w_incr carries the increment value, else it carries 0
   // If the decrement method is being invoked,
   //        w_decr carries the decrement value, else it carries 0

   (* fire_when_enabled, no_implicit_conditions *)
   rule r_incr_and_decr;
      value2 <= value2 + w_incr - w_decr;
   endrule

   // ---------------- METHODS

   method int read();
      return value2;
   endmethod

   method Action increment (int di);
      w_incr <= di;
   endmethod

   method Action decrement (int dd);
      w_decr <= dd;
   endmethod

endmodule: mkCounter_v2
```

endpackage: Tb

A.7.5   PulseWire

Source files are in directory: rwire/pulsewire

```
package Tb;

// ------------------------------------------------------------------
// A top-level module connecting the stimulus generator to the DUT

(* synthesize *)
module mkTb (Empty);

   Counter c1 <- mkCounter_v1;    // Instantiate a version 1 counter
   Counter c2 <- mkCounter_v2;    // Instantiate a version 2 counter (same interface type)

   // ----------------
   // Step 'state' from 1 to 10 and show both counter values in each state

   Reg#(int) state <- mkReg (0);

   rule step_state;
      if (state > 9) $finish (0);
      state <= state + 1;
   endrule

   rule show;
      $display ("In state %0d, counter values are %0d, %0d", state, c1.read (), c2.read ());
   endrule

   // ----------------
   // For counter c1, increment by 5 in states 0..6, decrement by 1 in states 4..10
   // The interesting states are 4..6

   rule incr1 (state < 7);
      c1.increment (5);
   endrule

   rule decr1 (state > 3);
      c1.decrement ();    // Decrement always by 1, so method has no argument
   endrule

   // ----------------
   // Same for counter c2 (increment by 5 in states 0..6, decrement by 1 in states 4..10)
   // The interesting states are 4..6
```

```
   rule incr2 (state < 7);
      c2.increment (5);
   endrule

   rule decr2 (state > 3);
      c2.decrement ();     // Decrement always by 1, so method has no argument
   endrule

endmodule: mkTb

// -----------------------------------------------------------------
// Interface to an up-down counter

interface Counter;
   method int read();                    // Read the counter's value
   method Action increment (int di);     // Step the counter up by di
   method Action decrement ();           // Step the counter down by 1
endinterface: Counter

// -----------------------------------------------------------------
// Version 1 of the counter

(* synthesize *)
module mkCounter_v1 (Counter);
   Reg#(int) value1 <- mkReg(0);  // holding the counter's value

   method int read();
      return value1;
   endmethod

   method Action increment (int di);
      value1 <= value1 + di;
   endmethod

   method Action decrement ();
      value1 <= value1 - 1;
   endmethod

endmodule: mkCounter_v1

// -----------------------------------------------------------------
// Version 2 of the counter

(* synthesize *)
module mkCounter_v2 (Counter);
   Reg#(int) value2 <- mkReg(0);  // holding the counter's value

   RWire#(int) rw_incr <- mkRWire();     // Signal that increment method is being invoked
   PulseWire   pw_decr <- mkPulseWire();  // Signal that decrement method is being invoked
```

```
// This rule does all the work, depending on whether the increment or the decrement
// methods, both, or neither, is being invoked
(* fire_when_enabled, no_implicit_conditions *)
rule doit;
    case (tuple2 (rw_incr.wget(), pw_decr)) matches
        { tagged Invalid,    False } : noAction;
        { tagged Valid .di, False } : value2 <= value2 + di;
        { tagged Invalid,    True  } : value2 <= value2 - 1;
        { tagged Valid .di, True  } : value2 <= value2 + di - 1;
    endcase
endrule

method int read();
    return value2;
endmethod

method Action increment (int di);
    rw_incr.wset (di);
endmethod

method Action decrement ();
    pw_decr.send ();
endmethod

endmodule: mkCounter_v2

endpackage: Tb
```

A.7.6   BypassWire

Source files are in directory: rwire/bypasswire

```
// Copyright 2010 Bluespec, Inc.  All rights reserved.

package Tb;

// -----------------------------------------------------------------
// A top-level module connecting the stimulus generator to the DUT

(* synthesize *)
module mkTb (Empty);

    Counter c1 <- mkCounter_v1;    // Instantiate a version 1 counter
    Counter c2 <- mkCounter_v2;    // Instantiate a version 2 counter (same interface type)

    // ----------------
    // Step 'state' from 1 to 10 and show both counter values in each state
```

```
   Reg#(int) state <- mkReg (0);

   rule step_state;
      if (state > 9) $finish (0);
      state <= state + 1;
   endrule

   rule show;
      $display ("In state %0d, counter values are %0d, %0d", state, c1.read (), c2.read ());
   endrule

   // ----------------
   // For counter c1, increment by 5 always, decrement by 2 in states 4..10

   rule incr1;
      c1.increment (5);
   endrule

   rule decr1 (state > 3);
      c1.decrement (2);
   endrule

   // ----------------
   // Same for counter c2 (increment by 5 always, decrement by 2 in states 4..10)

   rule incr2;
      c2.increment (5);
   endrule

   rule decr2 (state > 3);
      c2.decrement (2);
   endrule

endmodule: mkTb

// ----------------------------------------------------------------
// Interface to an up-down counter

interface Counter;
   method int read();                 // Read the counter's value
   method Action increment (int di);  // Step the counter up by di
   method Action decrement (int dd);  // Step the counter down by dd
endinterface: Counter

// ----------------------------------------------------------------
// Version 1 of the counter

(* synthesize *)
module mkCounter_v1 (Counter);
   Reg#(int) value1 <- mkReg(0);  // holding the counter's value
```

```
   method int read();
      return value1;
   endmethod

   method Action increment (int di);
      value1 <= value1 + di;
   endmethod

   method Action decrement (int dd);
      value1 <= value1 - dd;
   endmethod

endmodule: mkCounter_v1

// ----------------------------------------------------------------
// Version 2 of the counter

(* synthesize,
   always_enabled = "increment" *)
module mkCounter_v2 (Counter);
   Reg#(int) value2 <- mkReg(0);  // holding the counter's value

   Wire#(int) w_incr <- mkBypassWire();  // Signal that increment method is being invoked
   Wire#(int) w_decr <- mkDWire (0);     // Signal that decrement method is being invoked

   // --------------- RULES

   (* fire_when_enabled, no_implicit_conditions *)
   rule r_incr_and_decr;
      value2 <= value2 + w_incr - w_decr;
   endrule

   // --------------- METHODS

   method int read();
      return value2;
   endmethod

   method Action increment (int di);
      w_incr <= di;
   endmethod

   method Action decrement (int dd);
      w_decr <= dd;
   endmethod

endmodule: mkCounter_v2

endpackage: Tb
```

```
/* ================================================================
EXERCISE:

(1) Remove the attribute

        always_enabled = "increment"

    from the mkCounter_v2 module declaration, then compile with -verilog.
    Observe and explain the compiler warning message.

* ================================================================
*/
```

A.7.7 RWires and atomicity

Source files are in directory: rwire/rwire_atomicity

```
// Copyright 2010 Bluespec, Inc.  All rights reserved.

package Tb;

import FIFO::*;

// ----------------------------------------------------------------
// A top-level module connecting the stimulus generator to the DUT

(* synthesize *)
module mkTb (Empty);

   FIFO#(int) f <- mkPipelineFIFO;    // Instantiate a FIFO (see module def below)

   // ----------------
   // Step 'state' from 1 to 10

   Reg#(int) state <- mkReg (0);

   rule step_state;
      if (state > 9) $finish (0);
      state <= state + 1;
   endrule

   // ----------------
   // Enqueue and dequeue something on every cycle (in every state)

   rule e;
      let x = state * 10;
      f.enq (x);
      $display ("State %0d: Enqueue %0d", state, x);
```

```
        endrule

    rule d;
        let y = f.first ();
        f.deq ();
        $display ("State %0d: Dequeue %0d", state, y);
    endrule

    // ----------------
    // Also clear the FIFO in state 2

    rule clear_counter (state == 2);
        f.clear ();
        $display ("State %0d: Clearing", state);
    endrule

endmodule: mkTb

// ----------------------------------------------------------------
// A 1-element "pipeline" FIFO

(* synthesize *)
module mkPipelineFIFO (FIFO#(int));

    // STATE ----------------

    // The FIFO
    Reg#(Bool)        full      <- mkReg (False);
    Reg#(int)         data      <- mkRegU;

    RWire#(int)       rw_enq    <- mkRWire;                // enq method signal
    PulseWire         pw_deq    <- mkPulseWire;            // deq method signal

    Bool enq_ok = ((! full) || pw_deq);
    Bool deq_ok = full;

    // RULES ----------------

    // This rule does all the work, taking into account that 'enq()' and
    // 'deq()' may be called simultaneously when the FIFO is already full
    rule rule_update_final (isValid(rw_enq.wget) || pw_deq);
        full <= isValid(rw_enq.wget);
        data <= fromMaybe (?, rw_enq.wget);
    endrule

    // INTERFACE ----------------

    method Action enq(v) if (enq_ok);
        rw_enq.wset(v);
    endmethod
```

```
    method Action deq()   if (deq_ok);
       pw_deq.send ();
    endmethod

    method first()        if (full);
       return data;
    endmethod

    method Action clear();
       full <= False;
    endmethod

endmodule: mkPipelineFIFO

endpackage: Tb

/*
----------------------------------------------------------------
EXERCISE:

(1) Fix the above problem.

    HINT (one possible approach): in the 'enq()' method, set the
    'full' and 'data' registers.  In 'rule_update_final', just
    set 'full' to False whenever 'deq()' but not 'enq()' is invoked.
    In this case there would still be the choice of whether the rule
    or the clear method fired first; but this wouldn't matter any
    more, as they would both be assigning the same value to 'full'.


* ================================================================
*/
```

A.8   Polymorphism

Chapter examples are in directory: `polymorphism`

A.8.1   Polymorphic function

Source files are in directory: `polymorphism/function`

```
// Copyright 2010 Bluespec, Inc.  All rights reserved.

package Tb;

(* synthesize *)
module mkTb (Empty);
```

```
function td nop( td i );
   return i;
endfunction

function Bool equal( td i, td j ) provisos( Eq#(td) );
   return ( i == j );
endfunction

function td add2x( td i, td j ) provisos( Arith#(td) );
   return ( i + 2*j );
endfunction

function Bool i2xj( td i, td j ) provisos( Arith#(td), Eq#(td) );
   return ( (i - 2*j) == 0 );
endfunction

function Bool isBlock( td i ) provisos(Bits#(td,szTD));
   return (pack(i) & 3) == 0;
endfunction

function Bool isBlock2( td i ) provisos(Bits#(td,szTD),
                                        Add#(2,ununsed,szTD));  // 2 + unused == szTD
   return (pack(i) & 3) == 0;
endfunction

function Bool isMod3( td i ) provisos(Arith#(td),
                                      Eq#(td),
                                      Add#(4,ununsed,SizeOf#(td)));
   return (i % 3) == 0;
endfunction


/////////////////////////////////////////
rule show;
   $display(" nop is ", nop(10));

   Bit#(10) v1 = 2;
   $display(" equal %x %x ", equal(v1,v1), equal(v1,~v1));

   Bit#(8)  v2 = 2;
   Bit#(8)  v3 = 8;
   $display(" add2x = %x", add2x(v2,v3));

   $display(" i2xj  = %x", i2xj(v2,v3));

   $display(" isBlock = %x", isBlock(v2));

   $display(" isBlock2 = %x", isBlock2(v2));
```

```
      $display(" isMod3   = %x", isMod3(v3));

      // if you try to compile these two lines
      //    Bit#(2) v4 = 1;
      //    $display(" isMod3   = %x", isMod3(v4));

      // you get a compiler error, as the size of v4 is too small
      // to match the required provisos..

      $finish;
   endrule

endmodule

endpackage
```

A.8.2   Simple polymorphic function with provisos

Source files are in directory: polymorphism/provisos

```
// Copyright 2010 Bluespec, Inc.  All rights reserved.

package Tb;

// (* synthesize *)
module mkPlainReg( Reg#(tx) ) provisos(Bits#(tx,szTX));
   Reg#(tx) val <- mkRegU;

   return val;
endmodule

module mkPlainReg2( Reg#(Bit#(n)) );
   Reg#(Bit#(n)) val <- mkRegU;
   return val;
endmodule

///////////////////////////////////////////////////////////////////
// If you wrap your polymorphic module with a specific data type
// then you can synthesize it

(* synthesize *)
module mkPlainRegInt( Reg#(int) );
   Reg#(int) val <- mkPlainReg;

   method Action _write(int i) = val._write(i);
   method int    _read()       = val._read();
endmodule

module mkNotReg( Reg#(tx) ) provisos(Bits#(tx,szTX)
```

```
                                        ,Bitwise#(tx)  // comment this out to see error
                                        );
   Reg#(tx) val <- mkRegU;

   method Action _write(tx i) = val._write(i);

   method tx     _read();
      return ~val;  // Bitwise#(tx) added because of ~ here
   endmethod
endmodule

module mkDoubleReg( Reg#(tx) ) provisos(Bits#(tx,szTX),
                                         Arith#(tx));
   Reg#(tx) val <- mkRegU;

   method Action _write(tx i) = val._write(i);

   method tx     _read();
      return val * 2; // Arith#(tx) added here
   endmethod
endmodule


interface T2#(type ta, type tb);
   method Action  drive(ta ina, tb inb);
   method ta      result();
endinterface

module mkT2( T2#(tx,ty) ) provisos(Bits#(ty,sY),  // need to pack/unpack ty
                                   Arith#(tx),    // need to add with type "ta"
                                   Add#(SizeOf#(tx),0,SizeOf#(ty))  // must be the same size
                                   );
   Wire#(tx) val <- mkWire;

   method Action  drive(tx ina, ty inb);
      val <= ina + unpack(pack(inb));
   endmethod

   method tx      result();
      return val;
   endmethod
endmodule

interface T3#(type ta, type tb);
   method Action  drive(ta ina);
   method tb      result();
endinterface

module mkT3( T3#(ta,tb) ) provisos(Bits#(ta,sA),       // need to pack/unpack ta
                                   Bits#(tb,sB),
```

```
                                     Add#(ununsed,sB,sA), // needed by truncate
                                     Log#(sA,sB));
                                     // just for fun, make output roof of log2
   Wire#(tb) val <- mkWire;

   method Action  drive(ta ina);
      Bit#(sA) tmp  = pack(ina);
      Bit#(sB) tmp2 = truncate(tmp);
      val           <= unpack(tmp2);
   endmethod

   method tb       result();
      return val;
   endmethod
endmodule

(* synthesize *)
module mkTb (Empty);
   Reg#(Bool) r1 <- mkPlainReg;
   Reg#(int)  r2 <- mkPlainRegInt;

   Reg#(int)  r3 <- mkNotReg;
   Reg#(int)  r4 <- mkDoubleReg;

   T2#(UInt#(32),Int#(32)) modT2 <- mkT2;

   T3#(UInt#(32),UInt#(5)) modT3 <- mkT3;

endmodule

endpackage
```

A.9 Advanced types and pattern-matching

Chapter examples are in directory: `adv_types`

A.9.1 Enums

Source files are in directory: `adv_types/enum`

```
// Copyright 2010 Bluespec, Inc.  All rights reserved.


package Tb;

typedef enum { IDLE, DOX, DOY, FINISH } Tenum1 deriving(Bits, Eq);
```

```
typedef enum { S0=43, S1, S2=20, S3=254 } Tenum2 deriving(Bits, Eq);

(* synthesize *)
module mkTb (Empty);

   Reg#(int) step <- mkReg(0);

   Reg#(Tenum1) r1 <- mkReg( IDLE );

   Reg#(Tenum2) r2 <- mkReg( S3 );

   rule init ( step == 0 );
      $display("=== step 0 ===");
      $display("enum1 IDLE   => ", IDLE);
      $display("enum1 DOX    => ", DOX);
      $display("enum1 DOY    => ", DOY);
      $display("enum1 FINISH => ", FINISH);

      $display("enum2 S0 => ", S0);
      $display("enum2 S1 => ", S1);
      $display("enum2 S2 => ", S2);
      $display("enum2 S3 => ", S3);

      step <= step + 1;
   endrule

   rule step1 ( step == 1 );
      $display("=== step 1 ===");
      Tenum1 foo = IDLE;

      step <= step + 1;
   endrule

   rule step2 ( step == 2 );
      $display ("All done");
      $finish (0);
   endrule

   /////////////////////////////////////////
   rule watcher (step != 0);
   endrule

endmodule: mkTb

endpackage: Tb
```

A.9.2   Simple Processor Model

Source files are in directory: adv_types/simple_proc

```
// Copyright 2010 Bluespec, Inc.  All rights reserved.
package Tb;

import SimpleProcessor::*;

// The program encodes Euclid's algorithm:
//     while (y != 0) if (x <= y) y = y - x else swap (x <=> y)
// In this example, x = 15 and y = 27, so the output should be 3

InstructionAddress label_loop     = 3;
InstructionAddress label_subtract = 10;
InstructionAddress label_done     = 12;
InstructionAddress label_last     = 13;

Instruction code [14] =
    {
       tagged MovI { rd: R0, v: 0 },            // 0: The constant 0
       tagged MovI { rd: R1, v: 15 },           // 1: x = 21
       tagged MovI { rd: R2, v: 27 },           // 2: y = 27
       // label_loop
       tagged Brz { rs: R2, dest:label_done },  // 3: if (y == 0) goto done
       tagged Gt  { rd: R3, rs1: R1, rs2: R2 }, // 4: tmp = (x > y)
       tagged Brz { rs: R3, dest: label_subtract }, // 5: if (x <= y) goto subtract
       // swap
       tagged Minus { rd: R3, rs1: R1, rs2: R0 },  // 6: tmp = x;
       tagged Minus { rd: R1, rs1: R2, rs2: R0 },  // 7: x = y;
       tagged Minus { rd: R2, rs1: R3, rs2: R0 },  // 8: y = tmp;
       tagged Br  label_loop,                      // 9: goto loop
       // label_subtract
       tagged Minus { rd: R2, rs1: R2, rs2: R1 },  // 10: y = y - x
       tagged Br  label_loop,                      // 11: goto loop
       // label_done
       tagged Output R1,                           // 12: output x
       // label_last
       tagged Halt                                 // 13: halt
    };

// -----------------------------------------------------------------
// The testbench

(* synthesize *)
module mkTb (Empty);

   Reg#(InstructionAddress) ia <- mkReg (0);
   SimpleProcessor sp <- mkSimpleProcessor ();
```

```
    // Iterate through the instructions array, loading the program into
    // into the processor's instruction memory
    rule loadInstrs (ia <= label_last);
        sp.loadInstruction (ia, code [ia]);
        ia <= ia + 1;
    endrule

    // Start the processor executing its program
    rule go (ia == label_last + 1);
        sp.start();
        ia <= ia + 1;
    endrule

    // Wait till the processor halts, and quit
    rule windup ((ia > label_last + 1) && (sp.halted));
        $display ("Fyi: size of an Instruction is %0d bits", valueof (SizeOf#(Instruction)));
        $finish (0);
    endrule

endmodule: mkTb

endpackage: Tb



// Copyright 2010 Bluespec, Inc.  All rights reserved.
package SimpleProcessor;

// ------------------------------------------------------------------
// A small instruction set

// ---- Names of the four registers in the register file
typedef enum {R0, R1, R2, R3} RegNum
        deriving (Bits);

Integer regFileSize = 2** valueof (SizeOf#(RegNum));

// ---- Instruction addresses for a 32-location instruction memory
typedef 5 InstructionAddressWidth;
typedef UInt#(InstructionAddressWidth) InstructionAddress;
Integer imemSize =  2** valueof(InstructionAddressWidth);

// ---- Values stored in registers
typedef Bit#(32)  Value;

// ---- Instructions
typedef union tagged {
    struct { RegNum rd; Value v; }                 MovI;    // Move Immediate
    InstructionAddress                             Br;      // Branch Unconditionally
    struct { RegNum rs; InstructionAddress dest; } Brz;     // Branch if zero
    struct { RegNum rd; RegNum rs1; RegNum rs2; }  Gt;      // rd <= (rs1 > rs2)
```

```
   struct { RegNum rd; RegNum rs1; RegNum rs2; }   Minus;    // rd <= (rs1 - rs2)
   RegNum                                          Output;
   void                                            Halt;
} Instruction
  deriving (Bits);

// ----------------------------------------------------------------
// The processor model

interface SimpleProcessor;
   method Action loadInstruction (InstructionAddress ia, Instruction instr);
   method Action start ();     // Begin instruction execution at pc 0
   method Bool   halted ();
endinterface

(* synthesize *)
module mkSimpleProcessor (SimpleProcessor);

   // ---- Instruction memory (modeled here using an array of registers)
   Reg#(Instruction) imem[imemSize];
   for (Integer j = 0; j < imemSize; j = j + 1)
      imem [j] <- mkRegU;

   Reg#(InstructionAddress) pc <- mkReg (0);     // The program counter

   // ---- The register file (modeled here using an array of registers)
   Reg#(Value) regs[regFileSize];          // The register file
   for (Integer j = 0; j < regFileSize; j = j + 1)
      regs [j] <- mkRegU;

   // ---- Status
   Reg#(Bool) running <- mkReg (False);
   Reg#(UInt#(32)) cycle <- mkReg (0);

   // ----------------
   // RULES

   rule fetchAndExecute (running);
      let instr = imem [pc];
      case (instr) matches
         tagged MovI  { rd: .rd, v: .v }     : begin
                                                  regs[pack(rd)] <= v;
                                                  pc <= pc + 1;
                                               end
         tagged Br     .d                    : pc <= d;
         tagged Brz  { rs: .rs, dest: .d } : if (regs[pack(rs)] == 0)
                                                  pc <= d;
                                               else
                                                  pc <= pc + 1;
         tagged Gt    { rd:.rd, rs1:.rs1, rs2:.rs2 } : begin
```

```
                                      Bool b = (regs[pack(rs1)] > regs[pack(rs2)]);
                                      Value bv = extend (pack (b));
                                      regs[pack(rd)] <= bv;
                                      pc <= pc + 1;
                                   end
            tagged Minus { rd:.rd, rs1:.rs1, rs2:.rs2 } : begin
                                      regs[pack(rd)] <= regs[pack(rs1)] - regs[pack(rs2)];
                                      pc <= pc + 1;
                                   end
            tagged Output .rs                         : begin
                                      $display ("%0d: output regs[%d] = %0d",
                                      cycle, rs, regs[pack(rs)]);
                                      pc <= pc + 1;
                                   end
            tagged Halt                               : begin
                                      $display ("%0d: Halt at pc", cycle, pc);
                                      running <= False;
                                   end
         default: begin
                $display ("%0d: Illegal instruction at pc %0d: %0h", cycle, pc, instr);
                running <= False;
             end
      endcase
      cycle <= cycle + 1;
   endrule

   // ----------------
   // METHODS

   method Action loadInstruction (InstructionAddress ia, Instruction instr) if (! running);
      imem [ia] <= instr;
   endmethod

   method Action start ();
      cycle <= 0;
      pc <= 0;
      running <= True;
   endmethod

   method Bool halted ();
      return (! running);
   endmethod

endmodule: mkSimpleProcessor

endpackage: SimpleProcessor
```

A.9.3 Structs of registers vs. register containing a struct

Source files are in directory: adv_types/struct_reg

```
package Tb;

import FIFO::*;

typedef struct {
   int   a;
   Bool  b;
   } TEx1 deriving(Bits,Eq);

typedef struct {
   Reg#(int)  ra;
   Reg#(Bool) rb;
   } TEx2;

(* synthesize *)
module mkTb (Empty);

   Reg#(int) step <- mkReg(0);

   Reg#(TEx1)   r1 <- mkReg( unpack(0) );

   FIFO#(TEx1) fifo <- mkFIFO;

   TEx2 r2;
   r2.ra <- mkReg(2);
   r2.rb <- mkReg(True);

   Reg#(int)  reg3A <- mkReg(3);
   Reg#(Bool) reg3B <- mkReg(False);

   TEx2 r3;
   r3.ra = reg3A;
   r3.rb = reg3B;

   rule init ( step == 0 );
      $display("=== step 0 ===");
      $display("r1.a = ", r1.a);
      $display("r1.b = ", r1.b);

      TEx1 t = r1;
      t.a = 20;
      t.b = False;
      r1 <= t;
```

```
      step <= step + 1;
   endrule

   rule step1 ( step == 1 );
      $display("=== step 1 ===");
      $display("r2.ra = ", r2.ra);
      $display("r2.rb = ", r2.rb);

      r2.ra <= 31;
      r2.rb <= False;

      step <= step + 1;
   endrule

   rule step2 ( step == 2 );
      $display("=== step 2 ===");
      $display("r3.ra = ", r3.ra);
      $display("r3.rb = ", r3.rb);
      $display ("All done");
      $finish (0);
   endrule
endmodule: mkTb

endpackage: Tb
```

A.9.4   Tuples

Source files are in directory: adv_types/tuple

```
package Tb;

(* synthesize *)
module mkTb (Empty);
   Reg#(int) step <- mkReg(0);

   rule init ( step == 0 );
      $display("=== step 0 ===");

      Tuple2#( Bool, int ) foo = tuple2( True, 25 );

      Bool field1 = tpl_1( foo ); // this is value 1 in the list
      int  field2 = tpl_2( foo ); // this is value 2 in the list

      foo = tuple2( !field1, field2 );
```

```
        $display("tpl_1(foo) = ", tpl_1( foo ));
        $display("tpl_2(foo) = ", tpl_2( foo ));

        step <= step + 1;
    endrule

    rule step1 ( step == 1 );
        $display("=== step 1 ===");

        Tuple4#( Bool, UInt#(4), Int#(8), Bit#(12) ) bar = tuple4( True, 0, -2, 5 );

        $display("tpl_1(bar) = ", tpl_1(bar));
        $display("tpl_2(bar) = ", tpl_2(bar));
        $display("tpl_3(bar) = ", tpl_3(bar));
        $display("tpl_4(bar) = ", tpl_4(bar));

        step <= step + 1;
    endrule

    rule step2 ( step == 2 );
        $display("=== step 2 ===");
        $display ("All done");
        $finish (0);
    endrule

endmodule: mkTb

endpackage: Tb
```

A.10   Static elaboration - for loops/while loops

Chapter examples are in directory: `static_elab`

A.10.1   For-loops/While-loops

Source files are in directory: `static_elab/loops`

```
// Copyright 2010 Bluespec, Inc.  All rights reserved.

package Tb;

import FIFO::*;
import GetPut::*;
import ClientServer::*;

(* synthesize *)
```

```
module mkTb (Empty);

   Reg#(int) klimit <- mkReg (24);

   rule r;
      int a = 10;

      for (int k = 20; k < klimit; k = k + 1)
         a = a + k;

      $display ("rule r, a = %0d", a);
   endrule

endmodule: mkTb

endpackage: Tb

/*
----------------
EXERCISE:

- Replace the 'for' loop with a 'while' loop, like so:

      int k = 20;
      while (k < klimit) begin
         a = a + k;
         k = k + 1;
      end

   and you should see exactly the same behavior.

* ================================================================
*/
```

A.10.2 Static and dynamic conditionals

Source files are in directory: static_elab/conditional

```
// Copyright 2010 Bluespec, Inc.  All rights reserved.

package Tb;

import FIFO::*;
import GetPut::*;
import ClientServer::*;

Bool configVar = True;
int  x0        = 23;
```

```
(* synthesize *)
module mkTb (Empty);

    Reg#(int) cycle <- mkReg (0);

    int initval = (configVar ? (x0 + 10) : (x0 + 20));

    Reg#(int) r0 <- (configVar ? mkReg (initval) : mkRegU);

    Reg#(int) r1;
    if (configVar)
       r1 <- mkReg(initval);
    else
       r1 <- mkRegU;

    // ----------------
    // RULES

    rule r;
       $display ("%0d: rule r, r0 = %0d, r1 = %0d", cycle, r0, r1);

       r0 <= ((pack (cycle)[0]==1) ? (r0 + 1) : (r0 + 5));

       if (pack (cycle)[0]==1)
          r1 <= r1 + 1;
       else
          r1 <= r1 + 5;

       if (cycle > 7) $finish (0);
       cycle <= cycle + 1;
    endrule

endmodule: mkTb

endpackage: Tb
```

A.11 Expressions

Chapter examples are in directory: `expressions`

A.11.1 Don't care expressions

Source files are in directory: `expressions/dontcare`

```
package Tb;
```

```
(* synthesize *)
module mkTb (Empty);

   Reg#(int) step <- mkReg(0);

   rule init ( step == 0 );
      $display("=== step 0 ===");

      // for instance.. let's do the exact example above...
      bit o = 0;
      for (Bit#(2) i=0; i<2; i=i+1)
         for (Bit#(2) j=0; j<2; j=j+1) begin
            bit a = i[0];
            bit b = j[0];

            if ((a==0) && (b==0))
               o = 1;
            else
               o = 0;

            $display("%b%b = %b", a,b,o);
         end
      // this generates the NAND gate we expect and may be more natural
      step <= step + 1;
   endrule

   rule step1 ( step == 1 );
      $display("=== step 1 ===");

      bit o = 0;
      for (Bit#(2) i=0; i<2; i=i+1)
         for (Bit#(2) j=0; j<2; j=j+1) begin
            bit a = i[0];
            bit b = j[0];


            // but let's enumerate out the case we really care about
            if ( (a==0) && (b==0) )
               o = 1;
            else if ( (a==1) && (b==1) )
               o = 0;
            else
               o = ?;

            $display("%b%b = %b", a,b,o);
         end

      step <= step + 1;
   endrule
```

```
rule step2 ( step == 2 );
   $display("=== step 2 ===");

   bit o = 0;
   for (Bit#(2) i=0; i<2; i=i+1)
      for (Bit#(2) j=0; j<2; j=j+1) begin
         bit a = i[0];
         bit b = j[0];

         case ( { a, b } )
            2'b11: o = 1;
            2'b01: o = ?;
            2'b10: o = ?;
            2'b00: o = 0;
         endcase

         $display("%b%b = %b", a,b,o);
      end

   step <= step + 1;
endrule

rule step3 ( step == 3 );
   $display("=== step 3 ===");

   bit o = 0;
   for (Bit#(2) i=0; i<2; i=i+1)
      for (Bit#(2) j=0; j<2; j=j+1) begin
         bit a = i[0];
         bit b = j[0];

         case ( { a, b } ) matches
            2'b?1: o = ?;
            2'b10: o = 1;
            2'b00: o = 0;

         endcase

         $display("%b%b = %b", a,b,o);
      end

   step <= step + 1;
endrule

rule step4 ( step == 4 );
   $display("=== step 4 ===");

   $display ("All done");
   $finish (0);
```

```
      endrule

endmodule: mkTb

endpackage: Tb
```

A.11.2   Case expressions

Source files are in directory: expressions/case

```
// Copyright 2010 Bluespec, Inc.  All rights reserved.

package Tb;

(* synthesize *)
module mkTb (Empty);

   Reg#(int) step <- mkReg(0);

   Reg#(int) a <- mkReg(0);

   rule init ( step == 0 );
      $display("=== step 0 ===");

      int val = case( a )
                     0: return 25;
                     1: return 23;
                     2: return 17;
                     default: return 64;
                  endcase;

      $display("val = %b", val);

      step <= step + 1;
   endrule

   Bool initReg = True;

   Reg#(int) rr <- case ( initReg )
                     False: mkRegU;
                     True:  mkReg(0);
                   endcase;

   rule step1 ( step == 1 );
      $display("=== step 1 ===");
      $display("reg rr is initialized to ", rr );
      step <= step + 1;
   endrule
```

```
   rule step2 ( step == 2 );
      $display("=== step 2 ===");
      $display ("All done");
      $finish (0);
   endrule

endmodule: mkTb

endpackage: Tb
```

A.11.3   Function calls

Source files are in directory: `expressions/function_calls`

```
// Copyright 2010 Bluespec, Inc.  All rights reserved.

package Tb;

(* synthesize *)
module mkTb (Empty);

   function int square( int a );
      return a * a;
   endfunction

   function int increment( int a );
      // count up to maxInt then stop
      int maxInt = unpack({ 1'b1, 0 });
      if (a != maxInt)
         return a + 1;
      else
         return a;
   endfunction

   //////////////////////////////////////////////////////////////////
   Reg#(int) step <- mkReg(0);

   rule init ( step == 0 );
      $display("=== step 0 ===");
      $display("square    10 = ", square(10));
      $display("increment 20 = ", increment(20));

      $display("2*10+1       = ", increment(square(10)));

      step <= step + 1;
   endrule
```

```
   rule step1 ( step == 1 );
      $display("=== step 1 ===");
      $display ("All done");
      $finish (0);
   endrule

endmodule: mkTb

endpackage: Tb
```

A.12   Vectors

Chapter examples are in directory: `vector`

A.12.1   Arrays and square-bracket notation

Source files are in directory: `vector/arrays`

```
// Copyright 2008 Bluespec, Inc.  All rights reserved.
package Tb;


(* synthesize *)
module mkTb (Empty);
   UInt#(16) arr[10];

   for (Integer i=0; i<10; i=i+1)
      arr[i] = fromInteger(i * 3);

   UInt#(16) arr2[3][4];

   for (Integer i=0; i<3; i=i+1)
      for (Integer j=0; j<4; j=j+1)
         arr2[i][j] = fromInteger((i * 4) + j);

   Reg#(int) arr3[4];
   for (Integer i=0; i<4; i=i+1)
      arr3[i] <- mkRegU;

   rule load_arr3;
      arr3[0] <= 'h10;
      arr3[1] <= 4;
      arr3[2] <= 1;
      arr3[3] <= 0;
   endrule
```

```
   rule display_arr3;
      for (Integer i=0; i<4; i=i+1)
         $display("arr3[i] = %x", arr3[i]);
      $finish;
   endrule

endmodule: mkTb

endpackage: Tb
```

A.12.2   Arrays vs. Vectors

Source files are in directory: `vector/arr_vs_vector`

```
// Copyright 2008 Bluespec, Inc.  All rights reserved.

package Tb;

import Vector::*;
import FIFO::*;

(* synthesize *)
module mkTb (Empty);

   Int#(16) arr1[4];
   for (Integer i=0; i<4; i=i+1)
      arr1[i] = 0;

   Vector#(4, Int#(16)) vec1 = newVector;
   Vector#(4, Int#(16)) vec2 = replicate( 0 );

   Int#(16) arr2a[4];
   for (Integer i=0; i<4; i=i+1)
      arr2a[i] = 0;

   Vector#(4, Int#(16)) vec3 = map( fromInteger, genVector );

   Int#(16) arr3a[4];
   for (Integer i=0; i<4; i=i+1)
      arr3a[i] = fromInteger(i);

   function Int#(16) vec3init( Integer i );
      return fromInteger(i * 13);
   endfunction

   Vector#(4, Int#(16)) vec4 = map( vec3init,  genVector );

   Int#(16) arr4a[4];
```

```
   for (Integer i=0; i<4; i=i+1)
      arr4a[i] = fromInteger(i * 13);

   rule displayArray;
      for (Integer i= 0; i < 4; i=i+1)
         $display("arr1[i] = %x", arr1[i]);
   endrule

   rule displayVec2;
      for (Integer i= 0; i < 4; i=i+1)
         $display("vec2[i] = %x", vec2[i]);
   endrule


   rule displayVec3;
      for (Integer i= 0; i < 4; i=i+1)
         $display("vec3[i] = %d", vec3[i]);
   endrule


   rule displayVec4;
      for (Integer i= 0; i < 4; i=i+1)
         $display("vec4[i] = %d", vec4[i]);
   endrule

   rule finish;
      $finish;
   endrule

endmodule: mkTb

endpackage: Tb
```

A.12.3   bit-vectors and square bracket notation

Source files are in directory: `vector/bit_vector`

```
// Copyright 2008 Bluespec, Inc.  All rights reserved.

package Tb;

import Vector::*;

function Vector#(n,td) rotateUp( Vector#(n,td) inData, UInt#(m) inx)
   provisos( Log#(n,m) );
   return rotateBy( inData, inx );
endfunction
```

```
function Vector#(n,td) rotateDown( Vector#(n,td) inData, UInt#(m) inx)
   provisos( Log#(n,m) );
   UInt#(m) maxInx = fromInteger(valueof(n)-1);
   return rotateBy( inData, maxInx-inx );
endfunction

(* synthesize *)
module mkTb (Empty);
   rule test1;
      Bit#(5) foo1 = 5'b01001;
      $display("bit 0 of foo1 is ", foo1[0]);
      $display("bit 1 of foo1 is ", foo1[1]);
      $display("bit 2 of foo1 is ", foo1[2]);
      $display("bit 3 of foo1 is ", foo1[3]);
      $display("bit 4 of foo1 is ", foo1[4]);

      // You can also get a range of bits:
      Bit#(6) foo2 = 'b0110011;
      $display("bit 2:0 of foo2 is %b", foo2[2:0]);

      // and note that this returns a type Bit of the new size
      Bit#(3) foo3 = foo2[5:3];
      $display("bit 5:3 of foo2 is %b", foo3);

      // you can also set bits as you'd expect
      foo3 = 3'b111;
      $display("set foo3[2:0] to 1s => %b", foo3);
      foo3[1:0] = 0;
      $display("set foo3[1:0] to 0  => %b", foo3);

      Vector#(8, bit) bar1 = replicate(0);

      function bit isInteresting( Integer i );
         return (i % 3 == 0) ? 1 : 0;
      endfunction

      Vector#(16,bit) bar2 = genWith( isInteresting );
      $display("bar2 is %b", bar2);


      Vector#(16,bit) bar3 = unpack(16'b0101_0001_1000_1001);
      $display("bar3          is %b", bar3);

      $display("reverse bar3 is %b\n", reverse( bar3 ));

      $display("bar3          is %b", bar3);
      $display("rotate1 bar3 is %b", rotate( bar3 ));
      $display("rotate2 bar3 is %b\n", rotate(rotate( bar3 )));
```

```
        // rotate left or right doesn't necessarily make since
        // or at least I liked the idea of rotateUp and rotateDown
        // (you could of course wrap it in your own function with these)
        $display("bar3              is %b",   bar3);
        $display("rotate up 1 bar3 is %b",   rotateUp( bar3, 1 ));
        $display("rotate up 2 bar3 is %b",   rotateUp( bar3, 2 ));
        $display("rotate up 3 bar3 is %b\n", rotateUp( bar3, 3 ));

        $display("bar3              is %b",   bar3);
        $display("rotate dn 1 bar3 is %b",   rotateDown( bar3, 1 ));
        $display("rotate dn 2 bar3 is %b",   rotateDown( bar3, 2 ));
        $display("rotate dn 3 bar3 is %b\n", rotateDown( bar3, 3 ));

        // and you can still quickly pack and unpack them between a Bit
        Bit#(16) val = 25;
        bar3 = unpack( val );
        $display("bar unpacked = %b / %d", bar3, bar3);

        bar3[7] = 1;

        // and back
        Bit#(16) val2 = pack( bar3 );
        $display("bar packed   = %b / %d", val2, val2);
        $finish (0);
    endrule

endmodule: mkTb

endpackage: Tb
```

A.12.4   Whole register update

Source files are in directory: vector/whole_register

```
package Tb;

(* synthesize *)
module mkTb (Empty);
    Reg#(Bit#(8)) foo  <- mkReg(0);
    Reg#(Bit#(8)) foo2 <- mkReg(0);

    (* preempts = "up1, up2" *)

    rule up1;
        $display("foo[1:0] = %b", foo[1:0]);
```

```
    $display("foo[2]    = %b", foo[2]);

    foo[1] <= 1;

    let tmp = foo2;
    tmp[1] = 1;
    tmp[3] = 1;
    tmp[7:6] = 3;
    foo2 <= tmp;
endrule

rule up2;
    foo[2] <= 1;
endrule

Reg#(bit) b1 <- mkReg(0);
Reg#(bit) b2 <- mkReg(0);

rule up3;
    b1 <= 1;
    b2 <= 1;
endrule

rule done (b2 == 1);
    $display( "b12 = %b", {b1,b2} );
    $finish;
endrule

endmodule: mkTb

endpackage: Tb
```

A.12.5 Register containing a vector vs. vector of registers

Source files are in directory: vector/regvec_vecreg

```
package Tb;

import Vector :: *;

typedef enum {  OK,
                FAIL,
                TEST,
                WAIT
   } Status  deriving (Bits, Eq);

(* synthesize *)
module mkTb (Empty);
```

```
Reg#( Vector#(10,Status) ) regstatus <- mkReg( replicate(OK) ) ;
Vector#( 10, Reg#(Status) ) vecstatus <- replicateM (mkReg (OK)) ;

Reg#(Maybe#(UInt#(4))) failCond <- mkReg(Invalid);
Reg#(UInt#(4)) cycle <- mkReg(0);


rule checkChannel;
   cycle <= cycle + 1;
   if (cycle == 5)
      failCond <= tagged Valid 5;
   else
      failCond <= tagged Invalid;
   $display("cycle = ", cycle);
   if (cycle > 9) $finish(0);
endrule


rule updateRegStatus (failCond matches tagged Valid .chan );
   let tempstatus = regstatus;
   tempstatus[chan] = FAIL;
   tempstatus[chan + 1] = TEST;
   regstatus <= tempstatus ;
endrule

rule displayReg ;
    for (Integer i = 0 ; i < 10; i = i + 1) begin
      $display ("RegStatus(i)",  regstatus[i]);
      $display ("VecStatus(i)",  vecstatus[i]);
    end
endrule

rule updateVecStatus (failCond matches tagged Valid .chan );
   vecstatus[chan] <= FAIL ;
   vecstatus[chan + 1] <= TEST;
endrule

endmodule
endpackage
```

A.12.6   Vector of modules

Source files are in directory: `vector/modules`

```
// section 8 f - array of module
package Tb;

typedef Bit#(14) TD;

import FIFO::*;
import Vector::*;
import StmtFSM::*;

(* synthesize *)
module mkTb (Empty);

   // two fifos in
   Vector#(2,FIFO#(TD)) ififo <- replicateM( mkFIFO );

   // four fifos out
   Vector#(4,FIFO#(TD)) ofifo <- replicateM( mkFIFO );

   // if move0 fires, then move1 does not
   // this just gets rid of the warning you see otherwise
   // this is good enough for this example of index modules
   (* preempts = "move0, move1" *)
   rule move0;
      let data = ififo[0].first();
      ififo[0].deq();

      case ( data[13:12] )
         0: ofifo[0].enq( data );  // call (ofifo[0]) method enq( data )
         1: ofifo[1].enq( data );
         2: ofifo[2].enq( data );
         3: ofifo[3].enq( data );
      endcase
   endrule

   rule move1;
      let data = ififo[1].first();
      ififo[1].deq();
      case ( data[13:12] )
         0: ofifo[0].enq( data );
         1: ofifo[1].enq( data );
         2: ofifo[2].enq( data );
         3: ofifo[3].enq( data );
      endcase
   endrule

   // see more about test sequences in section 12
   Stmt test =
   seq
      // bits 13:12 select the output fifo
      ififo[0].enq( 'h0000 );
```

```
        ififo[0].enq( 'h1001 );
        ififo[0].enq( 'h2002 );
        ififo[0].enq( 'h3003 );
        ififo[1].enq( 'h0000 );
        ififo[1].enq( 'h1010 );
        ififo[1].enq( 'h2020 );
        ififo[1].enq( 'h3030 );
        noAction;
    endseq;
    mkAutoFSM ( test );


    // and now create 4 rules to watch each output using
    // static elaboration
    for (Integer i=0; i<4; i=i+1)
        rule watch;
            let data = ofifo[i].first();
            ofifo[i].deq();
            $display("ofifo[%d] => %x at ", i, data, $time);
        endrule

endmodule: mkTb

endpackage: Tb
```


A.12.7   Static elaboration

Source files are in directory: `vector/static_elab`

```
package Tb;

import Vector::*;

(* synthesize *)
module mkTb (Empty);
    Vector#(8,Reg#(int)) arr1 <- replicateM( mkRegU );
    Reg#(int) step <- mkReg(0);

    rule init ( step == 0 );
        for (Integer i=0; i<8; i=i+1)
            arr1[i] <= fromInteger(i);
        step <= step + 1;
    endrule

    rule step1 ( step == 1 );
        for (int i=0; i<8; i=i+1)
            arr1[i] <= i;
```

```
         step <= step + 1;
      endrule

      rule step2 ( step == 2 );
         for (int i=0;  i<8;  i=i+1)
            if (i != 7)
               arr1[i+1] <= arr1[i];
            else
               arr1[0]    <= arr1[7];
         step <= step + 1;
      endrule

      rule step3 ( step == 3 );
         for (int i=0;  i<8;  i=i+1)
            case (i)
               0: arr1[0] <= arr1[3];
               1: arr1[1] <= arr1[1];
               2: arr1[2] <= arr1[2];
               3: arr1[3] <= arr1[0];
               4: arr1[4] <= arr1[5];
               5: arr1[5] <= arr1[7];
               6: arr1[6] <= arr1[4];
               7: arr1[7] <= arr1[6];
            endcase
         step <= step + 1;
      endrule

      // now let's get run time...
      Reg#(UInt#(3)) inx <- mkReg(0);

      rule step4 ( step == 4 );
         arr1[inx] <= 0;
         step <= step + 1;
      endrule

      rule step5 ( step == 5 );
         for (Integer i=0;  i<8;  i=i+1) begin
            int indexVal = unpack(zeroExtend(pack(inx)));
         end
         step <= step + 1;
      endrule

      rule step6 ( step == 6 );
         for (Integer i=1;  i<8;  i=i+1)
            arr1[i-1] <= arr1[i] + fromInteger(i * 3);
         step <= step + 1;
      endrule

      rule step7 ( step == 7 );
         int acc = 0;
```

```
      for (Integer i=0; i<8; i=i+1) begin
         acc = arr1[i] + acc;
         arr1[i] <= acc;
      end

      step <= step + 1;
   endrule

   rule step8 ( step == 8 );
      $display ("All done");
      $finish (0);
   endrule

   rule watcher (step > 0);
      $display("=== step %d ===", step);
      for (Integer i=0; i<8; i=i+1)
         $display (" arr1[%d] = ", i, arr1[i]);
   endrule
endmodule: mkTb

endpackage: Tb
```

A.13   Finite State Machines

Chapter examples are in directory: fsm

A.13.1   Building a FSM explicitly using rules

Source files are in directory: fsm/rules

```
// Copyright 2010 Bluespec, Inc.  All rights reserved.
package Tb;

typedef enum { IDLE, STEP1, STEP2, STOP } State deriving(Bits,Eq);

// A top-level module connecting the stimulus generator to the DUT
(* synthesize *)
module mkTb (Empty);

   Reg#(State)    state <- mkReg(IDLE);
   Reg#(int)    counter <- mkReg(0);
   Reg#(Bool)   restart <- mkReg(False);

   rule runCounter;
      if (counter == 200) begin
```

```
            $display("Done");
            $finish;
         end
      counter <= counter + 1;
   endrule
   rule stateIdle ( state == IDLE );
      $display("Counter = %03d, State: IDLE", counter);

      // arc out of this state when counter = 4,8,12,etc
      if (counter % 4 == 0)
         state <= STEP1;
   endrule

   rule stateStep1 ( state == STEP1 );
      $display("Counter = %03d, State: STEP1", counter);
      if (restart)
         state <= IDLE;
      else if (counter % 8 == 0)
         state <= STEP2;
   endrule

   rule stateStep2 ( state == STEP2 );
      $display("Counter = %03d, State: STEP2", counter);
      state <= STOP;
   endrule

   rule stateSTOP ( state == STOP );
      $display("Counter = %03d, State: STOP", counter);
      state <= IDLE;
   endrule

endmodule

endpackage
```

A.13.2   One-hot FSM explicitly using rules

Source files are in directory: fsm/onehot

```
// Copyright 2010 Bluespec, Inc.  All rights reserved.
package Tb;

// -------------------------------------------------------------------
// A top-level module connecting the stimulus generator to the DUT

(* synthesize *)
module mkTb (Empty);
   Integer idle  = 0;
   Integer step1 = 1; // or "idle + 1",
```

```
   Integer step2 = 2;
   Integer stop  = 3;

   function Bit#(4) toState( Integer st );
      return 1 << st;
   endfunction

   Reg#(Bit#(4)) state <- mkReg( toState(idle) );

   // counter just to generate some conditions to move state along
   Reg#(int)      counter <- mkReg(0);
   Reg#(Bool)     restart <- mkReg(False);

   rule runCounter;
      if (counter == 200) begin
         $display("Done");
         $finish;
      end
     counter <= counter + 1;
   endrule

   (* mutually_exclusive = "stateIdle, stateStep1, stateStep2, stateStop" *)
   rule stateIdle ( state[idle] == 1 );
      $display("Counter = %03d, State: IDLE", counter);

      if (counter % 4 == 0)
         state <= toState( step1 );
   endrule

   rule stateStep1 ( state[step1] == 1 );
      $display("Counter = %03d, State: STEP1", counter);

      if (restart)
         state <= toState( idle );
      else if (counter % 4 == 0)
         state <= toState( step2 );
   endrule

   rule stateStep2 ( state[step2] == 1 );
      $display("Counter = %03d, State: STEP2", counter);
      state <= toState( stop );
   endrule

   rule stateStop ( state[stop] == 1 );
      $display("Counter = %03d, State: STOP", counter);
      state <= toState( idle );
   endrule

endmodule
```

```
endpackage
```

A.13.3   StmtFSM - Basic template

Source files are in directory: fsm/stmtfsm_basic

```
// Copyright 2010 Bluespec, Inc.  All rights reserved.
package Tb;

import StmtFSM::*;

(* synthesize *)
module mkTb (Empty);

   Reg#(Bool) complete <- mkReg(False);

   Stmt test =
   seq
      $display("I am now running at ", $time);
      $display("I am now running one more step at ", $time);
      $display("And now I will finish at", $time);
      $finish;
   endseq;

   FSM testFSM <- mkFSM (test);

   rule startit ;
      testFSM.start();
   endrule

   rule alwaysrun;
      $display("    and a rule fires at", $time);
   endrule

endmodule
endpackage
```

A.13.4   AutoFSM

Source files are in directory: fsm/autofsm

```
// Copyright 2010 Bluespec, Inc.  All rights reserved.
package Tb;

import StmtFSM::*;

(* synthesize *)
```

```
module mkTb (Empty);

   Stmt test =
   seq
      $display("I am now running at ", $time);
      $display("I am now running one more step at ", $time);
      $display("And now I will finish at", $time);
   endseq;

   mkAutoFSM ( test );

   rule alwaysrun;
      $display("   and a rule fires at", $time);
   endrule

endmodule

endpackage
```

A.13.5   StmtFSM example

Source files are in directory: fsm/example

```
package Tb;

import StmtFSM::*;
import FIFOF::*;

(* synthesize *)
module mkTb (Empty);

   FIFOF#(int) fifo <- mkFIFOF;
   Reg#(Bool)  cond <- mkReg(True);
   Reg#(int)     ii <- mkRegU;
   Reg#(int)     jj <- mkRegU;
   Reg#(int)     kk <- mkRegU;

   function Action functionOfAction( int x );
      return action
               $display("this is an action inside of a function, writing %d", x);
               ii <= x;
            endaction;
   endfunction

   function Stmt functionOfSeq( int x );
      return seq
               action
```

```
                        $display("sequence in side function %d", x);
                        ii <= x;
                    endaction
                    action
                        $display("sequence in side function %d", x+1);
                        ii <= x+1;
                    endaction
                endseq;
        endfunction

Stmt test =
seq
    // something of type "action" can be put on one line by itself
    $display("This is action 1"); // cycle 1
    $display("This is action 2"); // cycle 2
    $display("This is action 3"); // cycle 3

    // an action with several actions still only take one cycle
    action
        $display("This is action 4, part a");
        $display("This is action 4, part b");
    endaction

    // use functions to create actions that you are repeating a lot
    functionOfAction( 10 );
    functionOfAction( 20 );
    functionOfAction( 30 );

    // Create a function with my sequence and call it as needed
    functionOfSeq( 50 );

    // this whole action is just one cycle when it fires
    action
        $display("This is action 5");
        if (cond)
            $display("And it may have several actions inside of it");
    endaction

    // this is a valid "nop".. It's an action - it's noAction!
    noAction;

    repeat (4) noAction;

    repeat (4) action
                    $display("Repeating action!");
                    $display("Check it out");
                endaction

    seq
        noAction;
```

```
      noAction;
   endseq

   if ( cond ) seq
      // if cond is true, then execute this sequence
      $display("If seq 1");  // cycle 1
      $display("If seq 2");  // cycle 2
      $display("If seq 3");  // cycle 3
   endseq
   else seq
      // if cond is false, then execute this sequence
      action
         $display("Else seq 1");
      endaction
      action
         $display("Else seq 2");
      endaction
   endseq

   ////////////////////////////////////////
   // noice this is different that if/else inside a action
   // this takes one cycle when it executes
   action
      if (cond)
         $display("if action 1");
      else
         $display("else action 1");
   endaction

   action
      $display("Enq 10 at time ", $time);
      fifo.enq( 10 );
   endaction
   action
      $display("Enq 20 at time ", $time);
      fifo.enq( 20 );
   endaction
   action
      $display("Enq 30 at time ", $time);
      fifo.enq( 30 );
   endaction
   action
      $display("Enq 40 at time ", $time);
      fifo.enq( 40 );
   endaction

   ////////////////////////////////////////////////////////////
   // await is used to generate an implicit condition
   // await is an action which can be merged into another action
   // in this case I wanted to print a message when we were done
```

```
// while the fifo is notEmpty, the state machine
// sits here and no actions fire
action
   $display("Fifo is now empty, continue...");
   await( fifo.notEmpty() == False );
endaction

if ( cond )
   noAction; // A
else
   noAction; // B

for (ii <= 0; ii < 10; ii <= ii + 1) seq
   $display("For loop step %d at time ", ii, $time);
endseq

ii <= 0;

while (ii < 10) seq
   action
      $display("While loop step %d at time ", ii, $time);
      ii <= ii + 1;
   endaction
endseq


for (ii <=0; ii<=4; ii<=ii+1)
   for (jj <=0; jj<=5; jj<=jj+1)
      for (kk <=0; kk<=3; kk<=kk+1) seq
         action
            if (kk == 0)
               $display("kk = 000, and ii=%1d, jj=%1d", ii, jj );
            else
               $display("kk = --%1d, and ii=%1d, jj=%1d", kk, ii, jj );
         endaction
      endseq

par
   $display("Par block statement 1 at ", $time);
   $display("Par block statement 2 at ", $time);

   seq
      $display("Par block statement 3a at ", $time);
      $display("Par block statement 3b at ", $time);
   endseq

endpar
$display("Par block done!");
par
   seq
```

```
                ii <= 2;
                ii <= 3;
            endseq

            seq
                ii <= 1;
                ii <= 0;
            endseq

            ii <= 10;

        endpar

        $display("par block conflict test, ii = ", ii);
    endseq;
    FSM testFSM <- mkFSM( test );

    Stmt rcvr =
    seq
        while(True) seq

            action
                $display("fifo popped data", fifo.first());
                fifo.deq();
            endaction

            repeat (5) noAction;
        endseq
    endseq;

    FSM rcvrFSM <- mkFSM( rcvr );

    rule startit;
        testFSM.start();
        rcvrFSM.start();
    endrule

    rule finish (testFSM.done() && !rcvrFSM.done());
        $finish;
    endrule

endmodule

endpackage
```

A.14   Importing existing RTL

Chapter examples are in directory: `importbvi`

A.14.1 Verilog File

Source files are in directory: importbvi/verilog

```verilog
module SizedFIFO(V_CLK, V_RST_N, V_D_IN, V_ENQ, V_FULL_N, V_D_OUT, V_DEQ, V_EMPTY_N, V_CLR);
    parameter                   V_P1WIDTH = 1; // data width
    parameter                   V_P2DEPTH = 3;
    parameter                   V_P3CNTR_WIDTH = 1; // log(p2depth-1)
    // The -1 is allowed since this model has a fast output register
    parameter                   V_GUARDED = 1;

    input                       V_CLK;
    input                       V_RST_N;
    input                       V_CLR;
    input [V_P1WIDTH - 1 : 0]   V_D_IN;
    input                       V_ENQ;
    input                       V_DEQ;

    output                      V_FULL_N;
    output                      V_EMPTY_N;
    output [V_P1WIDTH - 1 : 0]  V_D_OUT;
endmodule
```

A.14.2 Using an existing interface

Source files are in directory: importbvi/FIFOifc

```
import FIFOF :: * ;


import "BVI" SizedFIFO =
module mkSizedFIFOtest #(Integer depth, Bool g) (FIFOF#(a))
//module mkSizedFIFOtest #(Integer depth) (FIFOF#(a))
        provisos(Bits#(a,size_a));

        parameter V_P1WIDTH = valueOf(size_a);
        parameter V_P2DEPTH = depth;
        parameter V_P3CNTR_WIDTH = log2(depth+1);
        parameter V_GUARDED = Bit#(1)'(pack(g));

        default_clock clk;
        default_reset rst_RST_N;

        input_clock clk (V_CLK)  <- exposeCurrentClock;
        input_reset rst_RST_N (V_RST_N) clocked_by(clk)  <- exposeCurrentReset;

        method enq (V_D_IN) enable(V_ENQ) ready(V_FULL_N);
```

```
        method deq () enable(V_DEQ) ready(V_EMPTY_N);
        method V_D_OUT first () ready(V_EMPTY_N);
        method V_FULL_N notFull ();
        method V_EMPTY_N notEmpty ();
        method clear () enable (V_CLR);

        schedule deq CF enq ;
        schedule enq CF (deq, first) ;
        schedule (first, notEmpty, notFull) CF (first,notEmpty, notFull) ;
        schedule (clear, deq, enq) SBR clear ;
        schedule first SB (clear, deq) ;
        schedule (notEmpty, notFull) SB (clear, deq, enq) ;
        schedule deq C deq;
        schedule enq C enq;

endmodule

(*synthesize*)
module mkTb();
   FIFOF#(Bit#(8)) i <- mkSizedFIFOtest(2, True);
endmodule
```

A.14.3   Defining a new interface

Source files are in directory: importbvi/GetPutifc

```
interface MyGetPut#(type t);
   interface Get#(t) g;
   interface Put#(t) p;
endinterface

import GetPut ::*;

import "BVI" SizedFIFO =
module mkSizedFIFOtest #(Integer depth, Bool guard) (MyGetPut#(a))
        provisos(Bits#(a,size_a));

        parameter V_P1WIDTH = valueOf(size_a);
        parameter V_P2DEPTH = depth;
        parameter V_P3CNTR_WIDTH = log2(depth+1);
        parameter V_GUARDED = Bit#(1)'(pack(guard));

        port V_CLR = 0;

        default_clock clk;
        default_reset rst_RST_N;

        input_clock clk (V_CLK)  <- exposeCurrentClock;
```

```
        input_reset rst_RST_N (V_RST_N) clocked_by(clk)   <- exposeCurrentReset;

        interface Get g;
            method V_D_OUT get () enable(V_DEQ) ready(V_EMPTY_N);
        endinterface

        interface Put p;
            method put(V_D_IN) enable(V_ENQ) ready(V_FULL_N);
        endinterface

        schedule g_get CF p_put ;
        schedule g_get C g_get;
        schedule p_put C p_put;

endmodule

(*synthesize*)
module mkTb();
  MyGetPut#(Bit#(8)) i <- mkSizedFIFOtest(2, True);
endmodule
```